Scenes from Deep Time

Scenes

Early Pictorial Representations of the Prehistoric World

from Deep Time

Martin J. S. Rudwick

The University of Chicago Press • Chicago and London

The University of Chicago Press, Chicago 60637
The University of Chicago Press, Ltd., London
© 1992 by The University of Chicago
All rights reserved. Published 1992
Paperback edition 1995
Printed in the United States of America

00 99 98 97 96 95   5 4 3 2

ISBN (cloth): 0-226-73104-9
ISBN (paper): 0-226-73105-7

Acknowledgments appear on pages ix–xiii.

Library of Congress Cataloging-in-Publication Data
Rudwick, M. J. S.
    Scenes from deep time : early pictorial representations of the
prehistoric world / Martin J. S. Rudwick.
        p.   cm.
    Includes bibliographical references and index.
    ISBN 0-226-73104-9
    1. Paleontology—History. 2. Science—History.   3. Man,
Prehistoric, in art.   I. Title.
QE705.A1R82   1992
560′.9—dc20   91-47677
                                                  CIP

Contents

Introduction

The dinosaurs are surely everyone's favorites at any natural history museum today. We see their reconstructed skeletons, often rearing above our heads; we see more lifelike reconstructions of their whole bodies, skin and all; and now we may even see them heaving from side to side, champing their jaws and grunting, thanks to the magic of modern robotics. More significantly, however, we also see them portrayed in their natural habitats, in relation to other animals—usually eating or being eaten—and to the plant life of the time, all set in a landscape scene; and any such realistic diorama is usually only one of a whole series of such scenes, portraying the broad sweep of the history of life on earth, all the way from the time of the earliest macroscopic organisms to the time of the first human beings.

It is easy to take these scenes from "deep time" for granted.[1] But they involve a conceptual and material construction of a very peculiar kind. Their realistic style invites us to imagine that we are seeing the deep past with our own eyes, unproblematically, as if from a time machine. Yet we also know that in fact they are based on extremely fragmentary evidence, fleshed out with a complex network of theoretical inferences. Furthermore, a moment's reflection shows that they are very far from any simple realism, not least because they crowd into one scene a variety of organisms that would be unlikely to pose together so obligingly in real life.

Such scenes from deep time are clearly a pictorial genre, as much ruled by visual conventions as any other. But artistic conventions do not fall ready-made from heaven, nor are they concocted or decreed at a given moment. They are the products of *historical* development; they are constructed in the course of artistic practice in specific historical circumstances. In this particular case, the practice is not only artistic but also scientific: it involves the intersection of two traditions, whether they are embodied in a single individual, shared in a collaboration between an artist and a scientist, or distributed within a larger team of scientists, artists, and technicians.

This book is about the first scenes from deep time. It traces the development of this pictorial genre from its earliest origins to the time when its visual conventions were well established. That makes it essentially a nineteenth-century story. The authors and artists of most of these scenes from deep time were well aware of earlier examples and often took them explicitly as models for their own work. A straightforward narrative is therefore the most effective way to trace the development of the genre, and the most appropriate framework for this book.

The book begins with a glance backward to an earlier tradition, namely that of biblical illustration, which arguably provided one important visual precedent for the new kind of scene (chapter 1). It then traces in detail the first attempts to construct scenes of the prehuman world on strictly natural evidence, drawing on visual precedents in natural history (chapter 2). It follows the gradually broadening repertoire of such scenes, as they became more familiar and more acceptable (chapter 3). The first full-scale sequence of scenes created such a significant precedent that it merits a chapter to itself (chapter 4). The narrative then traces the vast enlargement of public awareness of the prehuman world, not least as a result of the first full-scale three-dimensional models (chapter 5). The story ends around the 1860s, when scenes from deep time had come to be routinely deployed in a form and a style that are unmistakably continuous with those of the dioramas in our modern museums, and with their two-dimensional equivalents in our popular science books and television programs (chapter 6). Finally, beyond the end of the narrative, I review some of the implications of the story as a whole (chapter 7).

I have kept the narrative and interpretative commentary as brief as possible, in order to allow the greatest space and prominence to the scenes themselves. In fact, a major purpose of this book is simply to make these first scenes from deep time more widely available. Only a handful have ever been reproduced in modern times; many of the rest come from printed sources that may not be available even in major libraries; and a few were never fully published even at the time they were made.

As everyone knows, the visual impact of any picture is highly dependent on size: a movie made for the wide screen is just not the same on TV; Botticelli's *Venus* in even the finest coffee-table book is just not the same as in the Uffizi. So these first scenes from deep time are reproduced here ("Figures") with their original quality of detail and—so far as is practicable— at their original sizes. Those parts of the captions that are given in quotation marks are also original.

INTRODUCTION

The figures are accompanied by the written descriptions ("Texts") that their authors intended to explain the scenes. In most cases that intention is quite explicit; in a few, I have extracted from a longer text a passage that at least alludes to the author's understanding of the scene. These texts are reproduced with only minimal alterations for the sake of clarity; editorial additions are given in brackets, and omissions are indicated in the usual way. Translations into English are my own, except where otherwise noted. Notes immediately below some of the texts explain allusions that may not be clear to modern readers.

The endnotes and the list of sources for the figures and texts are primarily for readers who want to take this work further. They provide exact references and they document my interpretative commentary by referring to the work of other historians. Both primary and secondary sources are given in abbreviated form in the endnotes and source listings and in full in the bibliography.

I have not attempted to translate scientific terms, particularly the names of fossil animals and plants, into their modern equivalents. Paleontological readers will have no difficulty in identifying most of their familiar favorites and may enjoy puzzling over the rest. One purpose of this book is to encourage geologists, paleontologists, and museum scientists to reflect on the tacit dimension of the visual means by which they present their modern conclusions to a wider public. It is impossible to create scenes from deep time that embody no tacit message about the natural world; anyone who advises on or designs such scenes therefore does well to be aware of their implications, which may become clearer from considering their historical origins. Beyond such specialist circles, however, I hope the book will also be enjoyed by the proverbial general reader, and by ordinary visitors to our museums, as the story of how the depiction of any scenes from deep time first came to seem feasible at all.

Finally, of course, the book is also intended for my fellow historians of science. Many years ago, in an essay on the historical origins of geology's "visual language" of maps, sections, and other diagrams, I criticized the excessively textual orientation of scholarly work in all branches of the history of science, and I urged the need for greater analytical attention to pictorial sources.[2] The appeal seemed for several years to have fallen on deaf ears, but recently there have been welcome signs of interest—though more among sociologists than historians of science—in the role of visual materials in the making and propagation of scientific knowledge.[3] When a leading *philosopher* of science writes that "after our recent obsession with words it is well to reflect on pictures and carvings," there are real grounds for optimism![4] I hope that the juxtaposition of visual images and verbal texts in this book will encourage historians of science to begin to treat visual sources as seriously as they have always taken their textual sources, and indeed as routinely as the scientists they study did in the past, and still do today.

Acknowledgments

My interest in these scenes from deep time was first aroused many years ago when the late Jacques Roger—who as colleague and friend is deeply missed by many historians of science—came to give a lecture in Amsterdam and brought as a gift from Paris a copy of Louis Figuier's *La terre avant le Déluge*. I had never previously seen a copy of this nineteenth-century scientific bestseller. I was fascinated by its magnificent sequence of reconstructed scenes (see chapter 6) and intrigued by the close collaboration between artist and scientist that they represented. It was several years, however, before I was able to explore any further the origins of this pictorial tradition. The incentive to do so, for which I am grateful, came when Jim Moore invited me to contribute to a Festschrift of Darwinian essays for John C. Greene; my essay in that volume was in effect a first sketch of the interpretation embodied in this book.[5]

However, most of the scenes reproduced here come not from books such as Figuier's, which proudly advertised their pictorial riches, but from far more scattered sources such as the frontispieces of books with no other illustrations of the kind. The profoundly textual orientation of most historians of science—we need a word to express the visual and graphical equivalent of "illiteracy"—is such that with few exceptions these pictures have rarely been commented on in the secondary literature. To find them is like looking for the proverbial needle in a haystack, and I do not claim that my search has been exhaustive. However, the range of examples reproduced here would have been far poorer and more patchy had it not been for the collaboration of friends and colleagues around the world, who responded generously to my appeal for information on further examples of early scenes from deep time. For suggestions of this kind I am greatly indebted to Hugh Torrens (Keele), John Thackray (London), James A. Secord (London), William A. S. Sarjeant (Saskatoon), Rhoda Rappaport (Poughkeepsie), Goulven Laurent (Angers), Wolfhart Langer (Bonn), David Knight (Durham), Donald K. Grayson (Seattle), Gabriel Gohau (Paris), François Ellenberger (Paris), Adrian Desmond (London), Peter Bowler (Belfast),

Mike Bassett (Cardiff), and William B. Ashworth, Jr. (Kansas City). For valuable comments and information of other kinds, I am grateful to Simon Schaffer (Cambridge) and Jane R. Camerini (Madison); and to Mark Hineline, Philip Kitcher, and other members of the Science Studies Program at the University of California, San Diego.

I am deeply indebted to the Master and Fellows of Trinity College, Cambridge, for summer hospitality that made it possible to write a draft of this book in peace and quiet, far from an insistent telephone, close to the riches of the University Library, and with the incomparable inspiration of a view of Great Court. At La Jolla, I am similarly grateful to the Director and the Librarian at Scripps Institution of Oceanography, for giving me facilities in the library in which to complete the work (and an almost equally beautiful view of the Pacific).

The research was done with the support of grants from the National Science Foundation (grant no. SES-88-96206) and from the Academic Senate of the University of California, San Diego.

Photographs are reproduced by kind permission of Mrs. R. A. Gordon (fig. 42) and of the following institutions: University Library, Cambridge (figs. 1–7, 21, 22, 26–28, 30–34, 36, 37, 39–41, 43–56, 58–66, 105); Department of Geology, National Museum of Wales, Cardiff (fig. 20); Scripps Institution of Oceanography, La Jolla (fig. 35); British Library, London (figs. 9, 38, 57, 76–78, 82, 101, 104); Trustees of the British Museum, London (figs. 10, 11); Natural History Museum, London (figs. 67, 68, 70–75); University Museum, Oxford (fig. 19); Bibliothèque Centrale, Muséum Nationale d'Histoire Naturelle, Paris (figs. 12, 15). Other illustrations are from photographs of originals in my own possession, and of one diagram of my own design (fig. 106).

1

Creation and the Flood

A ny scene from deep time embodies a fundamental problem: it must make visible what is really invisible. It must give us the illusion that we are witnesses to a scene that we cannot really see; more precisely, it must make us "virtual witnesses" to a scene that vanished long before there were any human beings to see it.[1]

However, this problem is only slightly more acute than that faced by an artist depicting a historical scene in a similarly "realistic" style. Whether it comes from classical or biblical history—say, the Fall of Rome or the Fall of Babylon—the picture must make its viewers believe they are seeing a plausible representation of an event that neither they nor the artist have really witnessed at all. It must make them virtual witnesses of a scene that is reconstructed from the testimony of those who did see it. In the tradition of Western pictorial art, that testimony was overwhelmingly *textual* in character. Knowledge of material remains—for example, of the ruins of ancient Rome—could be used to supplement the texts; but the textual evidence from the classical or biblical authors remained paramount.[2] What they reported in words was translated by the artist into visual terms, according to the

1

pictorial conventions of the time and place in which the artist was working. What was judged to be a plausible or "realistic" representation was of course relative to those shared conventions.

It is hardly surprising that the earliest scenes that can be regarded in retrospect as being from "deep time" were firmly embedded in this artistic tradition of visual representations of scenes from the *human* past. In early modern Europe, scholars considered that the past history of the human race was recorded more or less fragmentarily in the chronicles of all literate societies. It was the task of the science of "chronology" to compare and evaluate these records critically, to correlate the various calendars by which they were dated, and to weld them all into a single universal history.[3] However, the chronologers of the seventeenth century found their task increasingly difficult, as they penetrated back in time beyond the records of ancient Greece and its temporal equivalents elsewhere. By the time they reached the Flood or Deluge, of which they believed they could detect at least some obscure testimony in the records of many ancient cultures, one such record outshone all others by its apparent clarity and detail. Of course the biblical record would have been given some privileged status anyway, because of its overarching religious role in the culture of Christendom; but it is important to recognize that most seventeenth-century scholars considered that it also deserved special attention on account of its value as *history*.

For the time before the Deluge, the record became even more obscure, though here too the early chapters of Genesis seemed to provide at least a bare outline of "antediluvian" characters and events. Finally, or rather, for the beginning of all things, the chronologers had to rely on the biblical narrative of Creation itself. By its very nature this could not be regarded as a human record of events, since even Adam had not been there to witness them until the sixth day of Creation. But the veracity of the record was only enhanced by its putatively divine origin.

This image of a world of limited time, in which Creation itself was not more than a few thousand years distant, was simply a part of taken-for-granted reality in early modern Europe. Like its spatial or astronomical counterpart, the "closed world" of the Ptolemaic cosmos, it was not adopted for reasons of religious prejudice, still less expressed in order to avoid ecclesiastical censorship. It embodied the generally agreed, apparently common-sense view of the world.

It was within this image of the world's history that the first scenes from relatively deep time came to be designed. In Western religious art, there was a long-standing tradition of depicting episodes such as Adam and Eve in the Garden of Eden and Noah's Ark riding out the Flood, as early scenes within much longer sequences.[4] In stained-glass windows or in tempera wall paintings, such cycles sought to represent visually, and thereby to make more accessible and persuasive, the Christian interpretation of cosmic history—all the way from the Creation recorded in Genesis, through the pivotal events of the life, death, and resurrection of Christ recorded in the Gospels, to the final Judgment foreshadowed in the Apocalypse. Traced from the medieval centuries into the art of the Renaissance and later periods, the pictorial conventions changed dramatically, but the program remained much the same. More significant was the invention of printing, especially the concurrent development of print-making, first in the form of woodcuts and later as copper engravings.[5] This made such pictorial cycles far more widely available: they could now be studied at home, at least in more affluent homes, within the covers of a book, rather than being seen only in the local church, or on a lifetime's pilgrimage to some more distant and more distinguished site.

For the purposes of this book, it is convenient to begin with a relatively late example of such cycles. The one chosen here is particularly appropriate, because it was masterminded by a man who was also a distinguished naturalist, and who possessed one of the finest collections of fossils in early eighteenth-century Europe. Johann Jacob Scheuchzer (1672–1733) was trained as a physician and spent most of his life in his native city of Zurich in a variety of positions that would now be regarded as broadly scientific. He traveled extensively in the Swiss Alps, at a time when exploring the more remote parts was still a hazardous undertaking, and he published voluminous works on the natural history of Switzerland. Like many naturalists at this time, he also had major interests in human history; and he published a history of his native country and edited a collection of relevant historical documents.

Those two areas of interest—natural history and civil history—came together in his work on fossils. For like many of his contemporaries, Scheuchzer believed that fossils were relics of the Deluge. They recorded the natural history of the country before the Deluge, but they also provided uniquely persuasive evidence for the reality of that distant historical event. Scheuchzer's *Herbarium of the Deluge* (*Herbarium Diluvianum*, 1709) depicted the wide range of fossil plants in his own collection, in a way that made it a valuable reference work long after his "diluvial" interpretation had been abandoned.[6] Puzzled by the total absence of human fossils, he later seized on a newly discovered specimen as being indeed that of "a man who was a witness of the Deluge" (*Homo Diluvii testis;* 1726); he did not live to witness its much later identification as a large amphibian![7]

4

Scheuchzer's scenes from near the beginning of time—as he and most of his contemporaries conceived it—were published in his last and largest work, *Sacred Physics* (*Physica sacra*, 1731–33). They came at the start of the sumptuous folio volumes, which were published in both Latin and French, the older and the newer international languages of science and scholarship, as well as in German, Scheuchzer's native language. Scheuchzer's work thereby became widely known throughout the literate world.[8] The word "physics" still bore its old Aristotelian meaning and was not far from the modern sense of "science." The work was "sacred" physics, because it sought to illustrate the biblical narrative from ancillary evidence drawn from the best science of the day. It was a massive undertaking. There were no fewer than 745 full-page copper engravings; indeed the German edition was even entitled *Copper Bible* (*Kupfer-Bibel*), in order to emphasize its illustrations. They were drawn by a team of eighteen engravers under the direction of the imperial engraver Johann Andreas Pfeffel (1674–1748), whose name was rightly given as much prominence on the title page as Scheuchzer's. Other artists had special responsibility for the design of the elaborate baroque frames for the scenes, and even for the lettering of the captions.

The overwhelming majority of Scheuchzer's scenes illustrate episodes from biblical history, stretching from beginning to end, from Genesis to the Apocalypse. Just as the Creation narrative was regarded as a brief prelude to the main—human—story of the world, so likewise the engravings that illustrate the Creation (and the later Deluge), several of which are reproduced here, come just from the start of a far longer sequence of scenes, covering in principle the whole of human history.

Unlike the literalism of modern fundamentalists, with their deliberate rejection of biblical scholarship, Scheuchzer's superficially similar interpretation of the

CHAPTER ONE

earliest chapters of the Bible reflects a mainstream tradition that in his day still embodied good plain sense. A
more historical understanding of Hebraic language and
imagery, theology and cosmology, as represented in early
work on biblical criticism, had not yet spread widely even
in scholarly circles. Scheuchzer and most of his contemporaries saw no difficulty in assuming that the Creation
and the Deluge had taken place just as and when a literal
reading of the texts suggested. That assumption is reflected visually in the engravings that he and Pfeffel designed to illustrate some scenes from the deepest time
they could imagine.

The most important feature of the scenes that illustrate the Creation narrative is that they form a *sequence*
that leads from initial chaos to a completed and human
world. The various components of the natural world, finally including mankind too, are brought in turn onto
the stage on which the drama of human redemption is to
be played out. However, although Scheuchzer himself
uses the traditional metaphor of the "theatre of the
world" (see text 4), the elaborate decorative frames to
these scenes suggest even more forcefully that they were
to be viewed as a sequence of pictures, set out as if along
the walls of a gallery, although in fact between the covers
of a book.

The first pictures, of initial chaos and the creation of
light, are depicted from a cosmic, not a terrestrial viewpoint—perhaps a divine view, but in any case certainly
not a human one. The selection reproduced here thus
starts with two scenes from the third day of Creation
(figs. 1, 2; text 1). They illustrate the world just before
and just after the creation of plant life. Before, the world
is bare and ugly, yet also like a well-tilled plant nursery,
ready and able to sustain a fertile world of plants. After, it
is lush, beautiful, and full of color. Yet—as Scheuchzer is
careful to add, in order to counter any suggestion of

6

Text 1
THE WORK OF THE THIRD DAY [OF CREATION]
In this Third Day, we see the surface of the Earth
raised up, the waters running down the slopes from
the hills, and the beds of seas, lakes and rivers filling
up. But the Earth is still quite naked and uniform,
with none of the ornament of a painting, and with its
dirty colour even inspiring a certain horror. However, this silt was a rich nursery; this muddy water
was at the same time pregnant and nourishing; the
Earth was fertile, so that plants of all kinds could
grow there; and in a moment it took on an attractive
verdure, enamelling the Earth with colours: but nevertheless without having in itself the power to produce everything.

Johann Scheuchzer, *Sacred Physics* (1731).
NOTE. This and the other texts from Scheuchzer's work
are translated from the French edition, which probably
reached a more widespread readership than either the
Latin or the German.

TAB. VI.

GENESIS Cap. I. v. 9. 10.
Opus tertiæ Diei.

I. Buch Mosis Cap. I. v. 9. 10.
Drittes Tagwerck.

Figure 1. "The Work of the Third Day": the creation of mountains, rivers and seas. "And God said, Let the waters under the heaven be gathered together unto one place, and let the dry land appear: and it was so. And God called the dry land Earth; and the gathering together of the waters called he Seas: and God saw that it was good." Engraving by Johannes Andreas Pfeffel, from Johann Scheuchzer, *Sacred Physics* (1731). In this and the following captions, the quotation from Genesis is of the verse cited in the original caption; the text is from the Authorized (or "King James") Version (1611), which was the standard English translation of the Bible at the time of Scheuchzer's work.

7

materialism—it is not the soil itself that has the power to produce all these varied plants, but God alone.

That same point is made in the text that explains the scenes illustrating the work of the fifth day of Creation (figs. 3, 4; text 2). As was usual among naturalists at the time, Scheuchzer sees in the diversity and marvelous adaptations of marine animals the primary evidence that they owe their existence to God's creative action rather than to any intrinsic power of the material elements. In the scene depicting fish and whales (fig. 3), the border is decorated with specimens of fish as if in an exhibit; in the scene with shellfish (fig. 4), the shells are likewise shown on land, stuck to a decorative arch of rock, rather than in their positions of life. Although such scenes purport to show episodes from the work of Creation in the deep past, their design reveals that they are just as much—or even more—a survey of the diversity of nature in the present, as it might be displayed in a museum.

With the sixth day of Creation (fig. 5; text 3), quadrupeds are added to the diversity of the world; or, equally, Scheuchzer's museumlike survey moves further up the traditional "scale of beings" toward man. The scene showing all these animals (others are depicted on another engraving not reproduced here) is drawn in a style that already had a long artistic history, one that continues to influence the genre of scenes from deep time today. The animals pose in a kind of tableau; one of each kind, hardly in interaction with each other or with their background of plant life. It is a scene of Arcadia, or of the Garden of Eden, lacking only the human presence.

That presence is provided in the scene that immediately follows (fig. 6; text 4). As Scheuchzer explains, all is now ready and prepared for the principal character to come on stage, for the host to sit down at table. All the previous phases of Creation have been merely prepara-

Text 2
THE WORK OF THE FIFTH DAY [OF CREATION]

God *created* fish; that is to say, they were not in any way produced by the power of the water itself. The only thing that the waters contributed was the place; the structure is the work of God. Do you wish to be persuaded of the truth of this statement? Consider on the one hand the simplicity of water; on the other, the admirable structure of the fish, which are differentiated in so many ways.

Johann Scheuchzer, *Sacred Physics* (1731).

Text 3
THE WORK OF THE SIXTH DAY [OF CREATION]

Birds, fish and insects have more similarity to Man than plants; quadrupeds, and reptiles of the serpent kind, approach still closer to Man than fish and birds. Thus we ascend by degrees from the structure of plants and animals to that of Man.

Johann Scheuchzer, *Sacred Physics*, (1731).

Text 4
THE WORK OF THE SIXTH DAY [OF CREATION]
(CONTINUED)

The most noble of all creatures, the Microcosm or epitome of all this great World, now makes his appearance in the Theatre of the World: now that the table is fully spread, the host can be seated. The Sun and the stars had first to shine; the atmosphere had to become pure, and proper for the respiration of plants and animals; the waters above had to be separated from the waters beneath, and the moist from the dry; the Earth had to be clothed with trees and shrubs, and ornamented with flowers and fruits; animals of all kinds had to be created. At last Man could appear, in order to be established as the Ruler of all the works of the hand of God.

Johann Scheuchzer, *Sacred Physics* (1731).

Figure 2. "The Work of the Third Day," continued: the creation of plants. "And God said, Let the earth bring forth grass, the herb yielding seed, and the fruit tree yielding fruit after his kind, whose seed is in itself, upon the earth: and it was so . . . and God saw that it was good." The design of the frame incorporates various botanical specimens. From Johann Scheuchzer, *Sacred Physics* (1731).

Figure 3. "The Work of the Fifth Day": the creation of fish. "And God created great whales, and every living creature that moveth, which the waters brought forth abundantly, after their kind, and every winged fowl after his kind: and God saw that it was good." From Johann Scheuchzer, *Sacred Physics* (1731).

Figure 4. "The Work of the Fifth Day," continued: the creation of shellfish. From Johann Scheuchzer, *Sacred Physics* (1731).

TAB. XXII

Figure 5. "The Work of the Sixth Day": the creation of quadrupeds. "And God said, Let the earth bring forth the living creature after his kind, cattle, and creeping thing, and beast of the earth after his kind: and it was so . . . and God saw that it was good." From Johann Scheuchzer, *Sacred Physics* (1731).

GENESIS Cap. I. v. 24. 25.
Opus sextæ Diei.

I. Buch Mosis Cap. I. v. 24. 25.
Sechstes Tagwerck.

Figure 6. "The Creation of Man from the Dust of the Earth (*Homo ex Humo*)." "And God said, Let us make man in our image, after our likeness: and let them have dominion over the fish of the sea, and over the fowl of the air, and over the cattle, and over all the earth, and over every creeping thing that creepeth upon the earth." From Johann Scheuchzer, *Sacred Physics* (1731).

13

tory to the coming of humanity in the person of Adam, who sits in the Garden of Eden, surrounded by the creatures he has been set there to rule, and gazing upward with awe toward the divine source of his unique nature and authority. But the illustrations decorating (if that is the right word!) the frame of the scene remind the viewer of man's complex fetal development, which, like the complexity of animal structure, is a sign of man's status as a merely created being, and perhaps also of his mortality.

There is nothing particularly original about the designs that Scheuchzer and Pfeffel devised to illustrate the Creation narrative. On the contrary, they drew on a rich artistic tradition of similar images.[9] Their sequence is reproduced here simply because it is representative of its time, and because it was widely known and admired. As this book will show, scenes like Scheuchzer's became in turn an important pictorial precedent for those based on a new source of "testimony" about the deep past, namely, fossils. This was a source that Scheuchzer never considered, because he regarded fossils as invaluable witnesses to another and later moment in relatively deep time.

After his scenes from the days of Creation, Scheuchzer's sequence moves on in traditional fashion through the drama of Adam and Eve in the Garden of Eden, and onward to Noah's building of the Ark and the coming of the Flood. The former gives him an excuse for illustrating fig leaves, serpents, and thorns; the latter, for displaying a wide range of fossils interpreted as relics of the Deluge. That interpretation also appears on the margins of one of his depictions of the Deluge (fig. 7; text 5). Here the door of the Ark is closed, ready to embark; would-be passengers without reserved seats are left stranded on land soon to be submerged.[10] In the frame of the scene are specimens of the fossils that,

Text 5
THE BEGINNING OF THE DELUGE

The time has at last come, when the family of Noah are to be delivered from the impious society of the rest of mankind, and for all the human species to be exterminated. . . . Among the innumerable relics of the Deluge that we collect and now carefully preserve, one finds many that clearly prove that the Deluge began in Spring, and more precisely in May. I have published elsewhere many from my own collection; here I show some more . . . "Here are new kinds of coins, the dates of which are incomparably more ancient, more important and more reliable than those of all the coins of Greece and Rome."

Johann Scheuchzer, *Sacred Physics* (1731).

NOTE. *New kinds of coins:* The quotation is from the review of Scheuchzer's *Herbarium of the Deluge* (1709), in the yearbook for 1710 published by the Academy of Sciences in Paris.

Figure 7. "The Beginning of the Deluge." "In the six hundredth year of Noah's life, in the second month, the seventeenth day of the month, the same day were all the fountains of the great deep broken up, and the windows of heaven were opened." From Johann Scheuchzer, *Sacred Physics* (1731). The figures in the frame are of fossils taken to be relics of the Deluge, and are identified as follows: I. an ear of barley; II. a hazelnut; III. a mayfly.

15

Figure 8. The Ark afloat on the subsiding waters of the Deluge. From Johann Scheuchzer, *Herbarium of the Deluge* (1709).

Scheuchzer believed, confirmed the exact season of the event. The connection between Deluge and fossils is in fact made more clearly in the small engraving that decorated the title page of Scheuchzer's earlier *Herbarium of the Deluge* (fig. 8). Here the Ark is seen riding on the subsiding waters of the Deluge, leaving some shells stranded on the shore in the foreground, ready to be preserved as fossils.

In the long run, however, Scheuchzer's assumption that all fossils originated at the Deluge was less important than his emphasis on their status as *witnesses* to a past event. To make this point, Scheuchzer borrowed the authority of the Academy of Sciences in Paris, quoting from a review of his own book, which had used the already commonplace analogy between fossils and the coins or medals of Greece and Rome (text 5). Just as the evidence of coins could supplement textual records in the reconstruction of the classical world, so fossils could supplement the still more scanty human records of the earliest periods of human history. That they might act as "testimony" to periods even deeper in time was not a possibility that would have occurred to Scheuchzer, or to most of his contemporaries, simply because they saw no reason to believe that time itself was significantly deeper than mankind.

During the later part of the eighteenth century, however, that possibility could no longer be ignored, at least by the naturalists who explored the thick piles of rock strata to be seen in sea-cliffs and mountainsides, and who collected distinctive sets of fossils from them, without ever finding any trace of human remains. These naturalists were understandably hesitant about putting any figure to the magnitude of time that might have been involved, because there was little on which to base any such conjecture. But the suspicion that man was a latecomer in a vastly older history of the world grew slowly to a near certainty. The book *Epochs of Nature* (*Les époques de la nature*, 1778), by the great French naturalist Georges, Count Buffon (1707–88), was particularly influential in this respect, even though it quickly became outdated in empirical terms. Buffon sketched a vast panorama of earth history divided into seven epochs—echoing or parodying the "days" of the Genesis narrative—in which mankind appeared only in the seventh and last.[11] Deeply *prehuman* time became for the first time conceivable.

It is therefore at first sight surprising that, almost a century after Scheuchzer's monograph on fossil plants, a similar illustrated book on fossils should have had as its frontispiece a scene that clearly looks back to Scheuchzer's little engraving of the Deluge and its fossils (fig. 9; compare fig. 8). James Parkinson (1755–1824) was a London physician and is best known for his masterly *Essay on the Shaking Palsy* (1817), which first clearly described the disease entity that now bears his name. In his spare time, however, Parkinson was a keen fossil collector. Early in

17

the new century, he set out to fill an obvious gap by writing a substantial book on fossils in English, which would appeal to the growing number of other collectors in Britain. His *Organic Remains of a Former World*, published in three quarto volumes over an eight-year period (1804–11), was richly illustrated with hand-colored copper engravings of fossils of many kinds.[12] However, the book was written in an already old-fashioned style as a series of diffuse "letters" to another collector, and the considerable impact of the work derived more from its fine illustrations than from its text.

Furthermore, the first volume (1804) showed that Parkinson was unaware of research on the Continent that was rapidly making his interpretation of fossils outdated, though by the third volume (1811) he had remedied this to some extent. Thus his frontispiece and the passage that alludes to it still imply that *all* the fossils he is going to describe and illustrate owe their origin to the Deluge (fig. 9; text 6). The scene that the London landscape painter Richard Corbould (1757–1831) designed for Parkinson does, however, modify Scheuchzer's design in a subtle but significant way. The Ark is now so small and distant that it can easily be missed at first glance; conversely, the shells—two of them unmistakably ammonites known only from the older strata—are more prominent. However, in his text Parkinson expresses a profound skepticism about all theories in geology, the "diluvial" theory presumably included. Certainly this conviction that theorizing should be replaced by rigorous fact-collecting characterized the group of enthusiasts, Parkinson among them, who soon afterward founded the Geological Society of London (1807). As the first body of its kind in the world, its suspicion of theorizing deeply influenced the self-consciously new science of geology in the early years of the nineteenth century.[13]

There was, however, an ulterior reason for this loudly

Text 6

To trace the operations of nature, in periods far behind all human record; to pronounce opinions respecting the structure and the inhabitants of a former world; and to endeavour to find out the ways of God in forming, destroying, and reforming the earth; do certainly appear to be tasks, to which the limited powers of man are but little adapted. But since the world we inhabit is evidently composed of the wrecks of a former world; the materials of which that world was composed are, of course, at hand for our examination. The remains too of its former inhabitants are frequently found preserved, in such situations as teach us something, not only respecting the extent of the changes, which have taken place on the surface of this globe; but even the particular element which was employed, as the chief instrument of destruction, and of renovation. Scripture, likewise, corroborated by the collateral evidence of all human tradition, supplies us with the grand leading facts; that after the complete formation and the peopling of this globe, it was subjected to the destructive action of an immense deluge of water; all the fountains of the great deep were broken up; the high hills that were under the whole heaven were covered; and every living substance was destroyed, which was on the face of the earth. Chemistry and mineralogy also furnish us with their aid, by which we are taught the several changes, of which these substances, under various circumstances, are susceptible.

By these aids, we may sometimes be enabled to form, perhaps, a tolerably correct judgment, respecting some of the grand changes which took place, during the vast revolution which this planet has experienced. But so very remote is the period, to which our minds are to revert: so loose, and so light, are the grounds on which our conjectures are to be built: and so great is the temptation to imagination, to take the place of judgment, that, among the several systems, of which I shall have occasion to make mention, you, not only, will hardly find one on which

Figure 9. The subsiding waters of the Deluge: the Ark is beached on the distant islet (at rainbow's end), and shells—including ammonites—are left stranded (in the foreground) to become fossils. Frontispiece, drawn by Richard Corbould, engraved by Samuel Springsguth, from James Parkinson, *Organic Remains of a Former World* (1804).

19

proclaimed empiricism, and perhaps also for Parkinson's choice of design for the frontispiece that purported to summarize his whole work. In the 1790s, as the Revolution in France lurched through its most violent phases, Parkinson had belonged to the London Corresponding Society, a pressure group for political reform that the authorities in England regarded as dangerously subversive. He seems later to have diverted his revolutionary fervor into the safer channels of humanitarian reform. Nonetheless, notions of the vast antiquity of the world were still widely regarded in England as suspiciously close to the advanced thinking that was held to have provoked the French Revolution. Parkinson's frontispiece, with its impeccably traditional appearance, may therefore have been designed in part to dispel any suspicion that his book on fossils had any subversive intentions, just as the Geological Society's later disavowal of theorizing certainly served to put the new science politically above suspicion.[14]

That methodological policy also deliberately encouraged a sharper disjunction between geological research and its use in biblical interpretation. On the one hand there was the burgeoning new science, restricting itself to a description and, at most, a low-level interpretation of phenomena, rejecting or discouraging ambitions to create any high-level or global "theory of the earth." On the other hand there was a continuing interest, in other sectors of educated society, in imagining the broad sweep of human history in traditionally Christian terms.

The heightened sense of the drama of human history associated with the Romantic movement was the context for a renewal, in the 1820s, of the pictorial tradition of depicting scenes from at least the deepest *human* time imaginable. Although initially this had no obvious connection with the new science of geology, it can be seen in retrospect to have provided an important precedent

you can venture to depend; but you will discover, that the majority, so far from possessing even probability, rather resemble the fictions of poets, than the reasonings of philosophers.

James Parkinson, *Organic Remains of a Former World* (1804).

for some of the earliest scenes from even deeper, *prehuman*, time.

By far the most important figure in this respect was the English painter John Martin (1789–1854).[15] As a young man out to make a career in London, Martin recognized that paintings of historical events were still the genre that could best lead to success and recognition in the art world. He rapidly developed a style so distinctive that in France—where his work was admired as much as in England—it even gave rise to the adjective *martinien*. Martin made a name for himself as a painter of what was termed "the sublime," integrating Neoclassical elements into a Romantic style that emphasized vastness of scale, dramatic atmospheric effects, and scenes of tragedy and destruction. He soon realized that even if his paintings sold they would never provide him with a decent living, so he also taught himself the craft of etching and printmaking. He learned how to exploit the technique of mezzotint to create engravings with newly subtle effects, and he took to publishing an edition of prints as soon as the exhibition of each major painting had demonstrated its popularity. In this way his work, at least in black and white, reached a far wider public than would ever see the original paintings.

The first of his paintings to create a sensation when it was exhibited in London was *The Fall of Babylon* (1819), a grandiose composition based not only on the biblical story but also on newly available illustrated accounts of the architecture of ancient Egypt and the Orient. *Belshazzar's Feast* (1821) was the first to be accompanied by an explanatory pamphlet, with an outline sketch enabling the painting to be "read" in every detail; it was also the first to be turned into a highly successful mezzotint edition. The following year, his vast panorama of the destruction of Pompeii by the eruption of Vesuvius in A.D. 79 showed how the same style could be used to illustrate secular as well as biblical history. But the painting most directly relevant to the theme of this book was Martin's *Deluge* (1826). This too was soon turned into a mezzotint (1828), with an explanatory pamphlet (fig. 10; text 7).

The composition is as traditional in detailed content as it is strikingly *martinien* in style. As in some of Martin's earlier paintings, the human drama is dwarfed by the natural elements. The Ark is even smaller than in Parkinson's scene (fig. 9), so small indeed that it can hardly be found without Martin's keyed sketch. It is placed on a 4,000-foot shelf on the slope of the 15,000-foot Mount Ararat—Martin was scrupulously precise about matters of scale—where it remains plausibly intact in the calm eye of the colossal storm raging around it. The mountains are collapsing in a tumble of huge boulders—unkind critics suggested that Martin had studied a load of coal being tipped into a coal-hole!—while human beings and wild animals alike try in vain to escape. But these animate elements are quite eclipsed by the inanimate, except where a flash of lightning throws into relief the edifying sight of a despairing blasphemer and his virtuous wife. Martin had no hesitation about adopting the traditional techniques of the earlier chronologers; he even did his own calculation to justify including the 969-year-old Methuselah among the unfortunate victims. The biblical inspiration was supplemented in the pamphlet by extracts from Byron's popular verse drama *Heaven and Earth* (1823), which likewise deals in Romantic style with the events around the Deluge.

There is nothing in this composition to suggest that Martin was in any way aware that the research of the geologists, both in Britain and on the Continent, had recently transformed scientific conceptions of the character of this Deluge event. In particular, there is no suggestion that the species of wild animals caught in the

Text 7
THE DELUGE

This representation of the Universal Inundation of the Earth, comprehends that portion of time when the Valleys are supposed to be completely overflowed, and the intermediate Hills nearly overwhelmed; and the people who have escaped from drowning there, are vainly flying to the mountains for safety, while others are crowding upon

"The rocky foreground—where await
Man, beast and bird their fearful doom."

Explanation [see outline sketch of fig. 10]

1. The sun, moon, and a comet in conjunction.
2. Noah's Ark, illumined amidst the general gloom by the last beams of the sun, and protected by the Omnipotent from the fury of the elements, raging both above and below it. The only part of the great waters which remains undisturbed is at the base of the rock, which sustains the Ark (marked No. 5).
3. Mountains bursting, or the fountains of the great deep breaking open—

"Rushing oceans every barrier rend."

4. Mount Caucasus, or Ararat—

"Shall yon exulting peak,
Whose glittering top is like a distant star,
Lie low beneath the boiling of the deep?
No more to have the morning sun break forth,
And scatter back the mists in floating folds
From its tremendous brow? no more to have
Day's broad orb drop behind its head at even,
Leaving it with a crown of many hues?
No more to be the beacon of the world,
For angels to alight on, as the spot
Nearest the stars?"

5. The horizontal line, marking the undisturbed portion of the waters.
6. Falling mountains, threatening instantaneous

Figure 10. *The Deluge,* mezzotint by John Martin (1828). The key to text 7 has been redrawn, because Martin's sketch does not show the numerals clearly enough for reproduction.

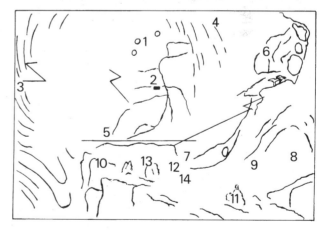

rising Deluge (fig. 11) were any other than those on view in the newly opened gardens of the Zoological Society in London and in the menagerie attached to the great Museum of Natural History in Paris. Only one feature is derived from the speculations of natural philosophers, and even that is an idea that had already been around for well over a century. In the hole left conveniently open in the background, Martin showed "the sun, moon and a comet in conjunction," thus alluding to a long-standing suggestion of a possible *natural* explanation for the cause of the Deluge.[16]

Martin later planned to illustrate the whole of the Bible, much as Scheuchzer had done a century before. That it remained incomplete, and a commercial failure, is certainly not due to waning public interest.[17] There is no need here to follow the tradition of biblical illustration any further, except to note that among the many artists influenced by Martin's work was the most famous of later illustrators of the Bible, Gustave Doré (1832–83).[18] Such illustrations continued their own career through the rest of the century; newer research, not only in geology but also in archaeology and biblical criticism, had surprisingly little impact on this highly traditional visual genre. By contrast, Martin's work was to have a profound effect on the pictorial representation of the truly deep past, as will become clear in chapter 3.

annihilation to the myriads of men and animals (No. 7) collected below, in the vain attempt to escape from the rising of the waters. The falling mountains are accompanied by the lightning and foaming torrents.

7. The myriads of people and animals crowded together.

8. An immense cavern, into which the multitudes attempt to rush in their hopeless search for safety from the falling mountains . . .

9. A den of ferocious animals.

10. Horsemen and others plunging in despair from a rock into the foaming deep.

11. A family, in silent depair, surrounded by howling wolves.

12. Lions and tygers, intermingled with human beings; but, in the great convulsion of nature, they harmlessly mix with men, and howl forth their prayer.

13. A blasphemer: his wife placing her hand upon his mouth to prevent his imprecations . . .

14. Methuselah.

John Martin, *A Descriptive Catalogue of the Engraving of the Deluge* (1828).

NOTES. *Their fearful doom:* a quotation from "Recollections of Mr Martin's Print of the Deluge," by the Quaker poet Bernard Barton (1784–1849), which Martin published with his own prose explanation. *Rushing oceans:* a phrase from Byron's poetic drama "Heaven and Earth" (1822); Martin reprinted lengthy extracts in his pamphlet explaining the picture. *Yon exulting peak:* another quotation from "Heaven and Earth"; the words are put in the mouth of Japhet, one of the sons of Noah.

The tradition of biblical illustration would serve as an important model for scenes from prehuman history. Pfeffel's engravings for Scheuchzer's *Sacred Physics* are an especially good example because they are relatively close in date and in artistic style to some of the earliest true scenes from deep time, and because Scheuchzer was also a major collector of the fossils that were to become

Figure 11. John Martin's *Deluge*:
detail of wild animals beyond the
human actors.

decisive evidence for the prehuman history of the earth. Scheuchzer himself regarded them instead as evidence for the biblical Deluge and accepted unquestioningly the traditional short time scale for the whole of cosmic history. But that is less important for our purposes than the fact that he, like other naturalists of his generation, treated fossils as natural "witnesses" to the past, analogous to the coins and monuments used by the historian of human time.[19]

In retrospect, perhaps the most significant feature of biblical illustrations such as Scheuchzer's was that they depicted a *sequence* of scenes in a temporal drama that had direction and meaning built into its structure. That model or precedent was therefore available to a later generation that sought to depict a comparable plot for far more ancient time and history.

Of all the traditional scenes depicting the earliest history of the world, those of the Flood survived into the early nineteenth century, least affected by the tides of Enlightenment rationalism, and most amenable to pictorial reinterpretation within the Romantic movement, as Martin's famous *Deluge* amply demonstrates. Again, it is only in retrospect that this can be seen to have affected the style and design of scenes from deeper, prehuman time. But it clearly imbued even the remote human past with a high sense of drama, grandeur, and mystery. Like the concept of a sequence of scenes, this Romantic aura was not unattractive to those who later sought to gain public support for the science on which a reconstruction of fully prehuman history could be based.

2

Keyholes into the Past

During the later eighteenth century, the suspicion that human history was dwarfed by pre-human history grew slowly to near certainty, at least among naturalists who concerned themselves with studying the topography of the earth's surface, the rocks of which it is composed, and the fossils that many of those rocks contain. It might seem surprising that this new conception of deep time was never translated into visual terms. But just what could these naturalists have portrayed in any imagined scene from deep time? Or, to put it another way, how would the content of any one scene have differed from another, or from a scene of the present-day world, if someone had tried to design it on purely *natural* evidence, to parallel scenes based on biblical and therefore *textual* evidence? The answer to these questions must be that such scenes could have had little content to distinguish them from scenes of the present world or from each other.

The new sense of the vast scale of the earth's history was based mainly on a growing awareness of the sheer thickness of the "formations" of rock strata that were often clearly exposed in sea-cliffs and mountainsides, and less dramatically in quarries, road-cuttings, and stream beds. Many of these strata looked as if they had been deposited quite slowly, particularly those that contained well-preserved shells, corals, and other fossils. In many areas, they appeared to abut rocks such as granite, which emerged from beneath the strata and rose up to form the cores of mountain chains. These "unstratified" rocks in mountain areas seemed to be the oldest materials exposed at the earth's surface, and accordingly were termed "Primary" or "Primitive." The strata that overlay them (though generally forming lower hills) were termed "Secondary": partly because they had clearly been formed at a later period, partly because some of them were composed of materials—granite pebbles, for example—evidently derived from Primary rocks. Down on the lowest ground were loose deposits such as river gravels and beds of sand and silt, which seemed to be derived in turn from Secondary (and Primary) rocks and were therefore termed "Tertiary." That tripartite division was thus a classification of rock types, but at the same time it was a crude periodization of earth history. (Another category, of "Transition" strata, was added later between the Primary and the Secondary, to accommodate the oldest formations with any trace of fossils.)[1]

A great variety of theories were put forward to account for this rough sequence of rock types. Some involved the idea of a hot primitive earth, cooling progressively through time. Others postulated a chemically complex fluid, initially global in extent, out of which a series of materials had been precipitated progressively, until the remnants came to form the present salty oceans. There was a conspicuous lack of consensus in all this theorizing: it was almost a matter of pride that every naturalist should have his own "theory of the earth." The important point is that none of them lent themselves to pictorial depictions of how the earth might have looked in the remote past to a time-traveling human observer: a hot ocean, or one precipitating the constituent crystals of granite, for example, would hardly have made for a striking picture. Not surprisingly, therefore, "theories of the earth" were illustrated visually, if at all, either by abstract diagrams or by pictures of present features—rock outcrops or striking specimens—that were regarded as decisive evidence for the theory being expounded: James Hutton's famous *Theory of the Earth* (1795) is a case in point.[2]

Naturalists were well aware that the Secondary strata contained fossils of kinds unknown in present seas: for example, ammonites, belemnites, and trilobites. In principle, these might have provided a basis for scenes from deep time, in which these organisms could have been depicted alive, in the manner of ordinary illustrations of present-day animals and plants. But so many strange new shells and other specimens were being brought back to Europe, by expeditions and voyages to remote parts of the world, that it seemed plausible to suppose that the fossils might represent organisms that were still alive somewhere on earth.[3] So an imagined scene from, say, the time when ammonites and belemnites were alive in European seas could not have differed significantly from a depiction of the "South Seas" or Pacific of the present day, with its Pearly Nautilus and other exotic organisms. Likewise, the most recent and "superficial" deposits, such as river gravels, contained, even in northern Europe, large bones that were commonly matched with those of elephants and other tropical animals. But this led simply to the inference—made, for example, by Buffon in his *Epochs of Nature* (1778)—that there had been a major climatic change in the course of earth history. That was certainly a striking conclusion; but an imagined scene from that remote time would not have differed significantly from a scene of, say, the jungles of India as they are known today.

This situation changed only at the very end of the

eighteenth century. The decisive factor was less the discovery of new evidence than the reassessment of material already available. Specifically, the well-known fossil bones just mentioned began to be studied with newly rigorous attention to their comparative anatomy. This led several naturalists to suspect, and then to insist, that the bones could not be matched with those of any living species, and that they must have belonged to species that were truly extinct. Once the reality of extinction was generally accepted, these fossils provided in principle the raw material for reconstructed scenes that would differ significantly from any scene drawn from the world of the present day. Unlike Scheuchzer's picture of the world just before the creation of man (fig. 5), it was now clear that any prehuman scene would have to feature some hitherto unknown animals. The fossil bones were mute yet eloquent "witnesses" of a world for which there was no evidence from the traditional *textual* sources.

Several naturalists became convinced around the same time that these fossil bones were of extinct species. But it was the work of Georges Cuvier (1769–1832) that caught the imagination of the whole educated world. Determined to make a career in Paris just after the Revolution, Cuvier carved a niche for himself at the reformed Museum of Natural History, the world center for this kind of science. His main research area of comparative anatomy gave him the key with which to resolve decisively the identity of the fossil bones.[4]

Soon after Cuvier started work at the museum, a striking example landed on his desk. A set of engravings of a fossil mammal was brought to Paris, and it fell to Cuvier to report on it. The bones had been sent to Spain from South America in 1789, and the drawings came from the Royal Museum in Madrid, where Juan Bautista Bru (1740–99) had reassembled the separated bones into an almost complete skeleton mounted in a somewhat stiff

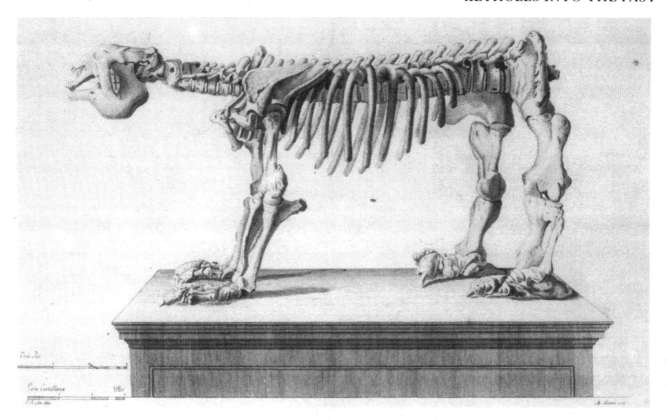

Figure 12. Juan Bautista Bru's reconstruction of the skeleton of an extinct mammal from South America, as mounted in the Royal Museum in Madrid, reproduced from Georges Cuvier's copy of the original print in the Museum of Natural History, Paris. This is the first of a set of six plates (the others are of individual bones) drawn by Bru and engraved by Manuel Navarro.

and awkward pose (fig. 12).[5] Based on a rare find of a virtually complete set of fossil bones, this was probably the first reconstruction of its kind. Such reconstructions of the skeletons of fossil vertebrates soon became commonplace, however, first in Cuvier's publications and then more generally. Eventually—though this is to jump ahead in time—they became as a matter of routine the first stage in the reconstruction of complete prehistoric scenes.

In style, Bru's engraving belongs to a pictorial tradition as old as comparative anatomy itself: a strictly lateral profile drawing provided the most effective visual summary of almost any animal. In this way, the pictorial traditions of comparative anatomy, and of natural history in general, became as important as those of biblical illustration in providing precedents for the new genre of scenes from deep time.

Cuvier himself did not attempt to depict more than the skeleton itself. In this case, he concluded that "the Paraguay animal" was a giant sloth, which he named *Megatherium* ("huge beast"); since no such creature had been reported live, he inferred that it was extinct.[6] Emboldened by this example, he then went on to study the well-known fossil elephant bones of northern Europe, and he concluded that they belonged to an extinct subarctic species, the mammoth. Many other spectacular mammals followed in quick succession, until Cuvier seemed to have assembled a virtual zoo of extinct animals, to match the living ones in the museum's menagerie just round the corner from his house. His skeleton of the *Mastodon* (fig. 13), an extinct elephant from North America, typifies his lifelike reconstructions—far more persuasive than Bru's awkward megatherium—while the vegetation beneath its feet provides at least a minimal suggestion of an environment.

At the same time, Cuvier was studying the fossil bones

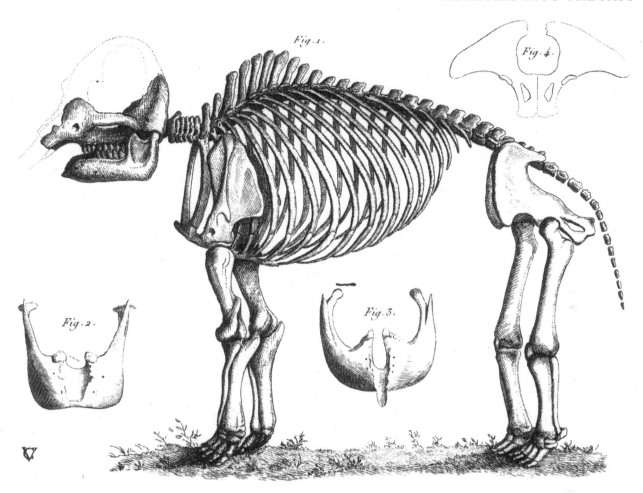

Figure 13. Georges Cuvier's reconstruction of the skeleton of the extinct elephantlike *Mastodon*, based on fossil bones found in Ohio (1806).

from the evidently much older Tertiary strata at Montmartre (then on the outskirts of Paris), which were mined for building stone and for the gypsum used to make "plaster of Paris." These bones posed a much tougher challenge; they were not only disarticulated and dispersed, but also much less like those of any living animals. By detailed anatomical study, however, Cuvier was able to reconstruct the fairly complete skeletons of two distinct genera, both quite unlike any living mammals. He named them *Palaeotherium* ("ancient beast") and *Anoplotherium* ("unarmed beast") and distinguished two species of each.[7] Once again, his reconstructed skeletons are persuasively lifelike (fig. 14).

Cuvier then took a significant step toward constructing a true *scene* from deep time. In a series of drawings that are superb examples of his artistic skill, he reconstructed the likely form of the major muscles attached to the skeleton, clothed them with skin to form a body outline and even, more conjecturally, inferred such unpreserved features as the eyes and ears (fig. 15). He had long made museum specimens his primary research material, and dissection his primary method; but these lively drawings reveal the depth of his understanding of the living animal body and must surely be based on intensive study of living mammals in the museum's menagerie.

The drawings are undated and may not have been made until after Cuvier collected all his research articles together and reprinted them in four impressive quarto volumes as his *Researches on Fossil Bones* (*Recherches sur les ossemens fossiles*, 1812). A decade later, he published an enlarged and much improved edition (1821–24), in which the Montmartre mammals were analyzed in greater detail and illustrated with an even larger number of finely engraved plates based on Cuvier's drawings of the bones.[8] In the text, he gave a vivid description of each animal, inferring not only its appearance but also its

Text 8

PALAEOTHERIUM MINUS

If we were able to bring this animal back to life as easily as we have reassembled its bones, we would think that what we were seeing running was a tapir smaller than a roe deer, with light and spindly limbs. That was certainly its appearance. Its height at the withers would have been sixty to seventy inches.

PALAEOTHERIUM MAGNUM

This animal was four and a half feet or more in height at the withers; that is the size of the rhinoceros of Java. It was lesser in height than a large horse, more squat, with a more massive head, stouter and shorter limbs, etc. There is nothing easier than to represent it in the living state.

ANOPLOTHERIUM COMMUNE

It was of considerable height at the withers; it could reach more than three feet and a few inches. But what distinguished it most was its long tail. This gave it something of the appearance of the otter, and it is very probable that like that carnivore it was often on or under the water, above all in marshy places; but this was certainly not in order to fish there. Like the water rat, the hippopotamus, and all kinds of boar and rhinoceros, our Anoplotherium was a herbivore; so it went in search of the succulent roots and stems of aquatic plants. With its habits as a swimmer and diver, it would have the smooth skin of the otter, or even perhaps the semi-bare skin of the pachyderms we are about to discuss. It is unlikely that it would have had long ears, which would have impeded it in its aquatic mode of life, and I can readily imagine that in this respect it resembled the hippopotamus and other quadrupeds that live mostly in water.

ANOPLOTHERIUM GRACILE

It would be a little more than two feet in height at the withers; equalling the chamois, although less

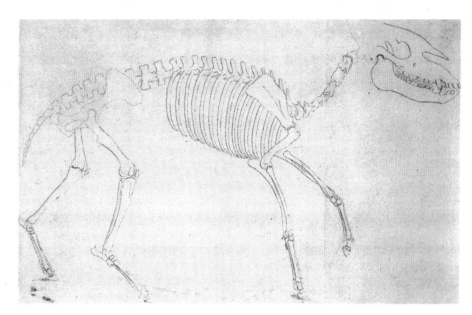

Figure 14. Georges Cuvier's reconstruction of the skeleton of the extinct mammal *Palaeotherium* (1808), based on fossil bones found in the Tertiary strata of Paris.

Figure 15. Georges Cuvier's unpublished reconstruction of the extinct mammal *Anoplotherium commune,* with body outline and musculature inferred from the reconstructed skeleton, based in turn on fossil bones found in the Tertiary strata of Paris.

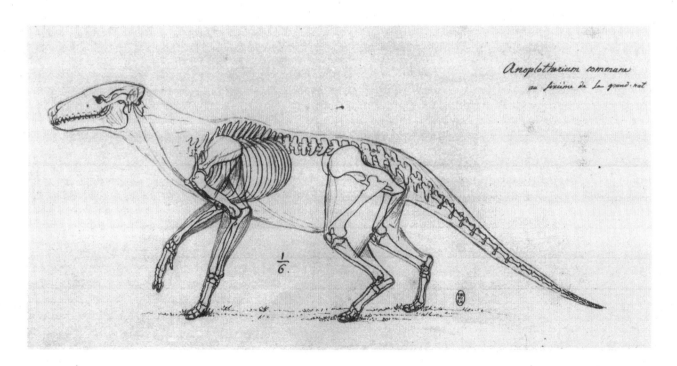

habits of life (text 8). Together they add up to an impressive evocation of a whole landscape and of the ecology of its fauna.

However, Cuvier chose not to publish his own superb manuscript reconstructions along with his text. The images that appear in the book fall far short of Cuvier's lively prose. He evidently delegated to his assistant Charles Leopold Laurillard (1783–1853) the task of drawing mere body outlines of the four species (fig. 16). Laurillard's artistic skills were far inferior to Cuvier's, and his reconstructions are so stiff and stylized that they recall nothing so much as the figures on the friezes from ancient Egypt, which were being published in Paris at just this time. In the rest of the seven volumes, there were no reconstructions at all, beyond those of plain skeletons.

The reason for this hesitancy about *publishing* pictorial scenes from deep time probably lies in Cuvier's concern for the scientific status of his work on fossil animals. To safeguard his scientific prestige and authority, and hence also his career, he needed to ensure that no one could criticize it as "mere speculation."[9] He had to demonstrate that every detail of his comparative anatomy of fossils was firmly grounded in a rigorous analysis of bones that were publicly available in the museum. Reconstructions of skeletons—and there were not many even of those—were as far as he was generally prepared to go. Beyond that, he chanced his reputation only with this one set of reconstructed body outlines (fig. 16), and even that bore his assistant's name, not his own.[10]

In the same year that Cuvier brought out the volume in which those minimal reconstructions appear, his most prominent English follower delivered a major paper to the Royal Society in London, which incorporated a more detailed and more sensational reconstruction of another scene from deep time. But again, this was textual rather than visual in character. In 1821, William Buckland

massive in head and bones; but this is owing to the great elongation of its limbs; its head was scarcely equal to that of the antelope. Whereas the Anoplotherium commune would have been heavy and shuffling in appearance when it walked on the ground, the [A.] gracile would have been agile and graceful. As light as the gazelle or roe deer, it would have run rapidly around the marshes and ponds in which the former species swam. It would have grazed on the aromatic plants of the dry ground, or on the shoots of the shrubs. Its movement was doubtless not hampered at all by a long tail. Like all agile herbivores, it was probably a timid animal; and large, highly mobile ears, like those of stags, would warn it of the least danger. Finally, its body was doubtless covered with a short coat of hair, and consequently we lack only its colour, in order to paint it as it formerly enlivened this countryside [around Paris], where it has been possible, after so many centuries, to unearth its scanty remains.

To complete the conception that one can make of this ancient animal population, we have represented several species, at their proportional sizes, and with the external forms that we have attributed to them [fig. 16].

Georges Cuvier, *Researches on Fossil Bones,* second edition (1822).

NOTE. *So many centuries:* Cuvier is here prudently cautious about quantifying the geological time scale, since any figure was bound to be highly speculative. He was well aware, from his stratigraphical work with his colleague Alexandre Brongniart (also published in his *Researches*), that even the Paris strata were far older than the superficial gravels with mammoths, etc., which in turn he believed to be prehuman and extremely ancient. Elsewhere, he referred quite casually to "thousands of centuries" as the likely age of the Paris strata.

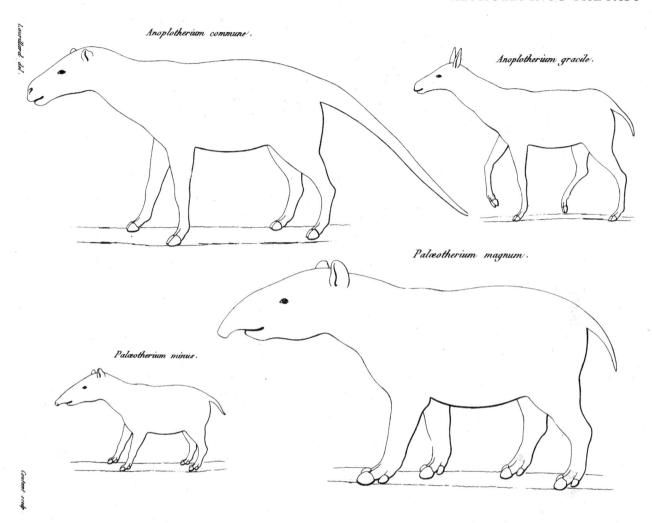

Figure 16. Reconstructed body outlines of the extinct mammals *Palaeotherium* and *Anoplotherium* from the Tertiary strata of Paris, drawn by Cuvier's assistant Charles Laurillard, engraved by Jean Louis Denis Coutant, and published in the second edition of Georges Cuvier's *Researches on Fossil Bones* (1822).

37

(1784–1856), the professor of geology at Oxford, was informed of a small cave that had recently been discovered near the village of Kirkdale in Yorkshire, from which a large number of fossil bones had already been collected. He carried out a thorough study of the cave and its bones and decided that they pointed to a conclusion quite different from the one he had anticipated. The bones, he inferred, had not been swept into the cave by any kind of Deluge, which was an event that he and most other geologists believed real enough, though enigmatic in character and cause. Instead they had been taken in as food by a pack of hyenas, who had used the cave as a den. This reconstruction was of a scene *before* the Deluge event, which itself had merely wiped out the fauna and brought the occupation of the cave to an end. Thereafter the bones in their muddy sediment had been sealed in by the slow accumulation of stalagmite (text 9).[11]

Buckland's fine verbal reconstruction certainly reflects a growing sense among geologists that it was both possible and legitimate to make explicitly *historical* inferences about the deep past. But it would hardly deserve inclusion in this book had it not evoked from one of Buckland's colleagues a response that is no less important for being humorous. William Daniel Conybeare (1787–1857) had been a young don at Oxford with Buckland during the early years of the Geological Society. Later he had been ordained and had moved to country parishes, first near Bristol and later near Cardiff; but like many other parson-naturalists, he had continued his scientific research unabated. In the same year as Buckland's paper on Kirkdale cave, Conybeare published his *Outlines of the Geology of England and Wales* (1822), which quickly became the authoritative reference book on the stratigraphical succession, making his name well known to geologists throughout the scientific world.[12]

Conybeare's comment on Buckland's cave research

Text 9

It must already appear probable, from the facts above described, particularly from the comminuted state and apparently gnawed condition of the bones, that the cave at Kirkdale was, during a long succession of years, inhabited as a den of hyaenas, and that they dragged into its recesses the other animal bodies whose remains are found mixed indiscriminately with their own: this conjecture is rendered almost certain by the discovery I made, of many small balls of the solid calcareous excrement of an animal that had fed on bones. . . . It was at first sight recognised by the keeper of the Menagerie at Exeter Change, as resembling, both in form and appearance, the faeces of the spotted or Cape hyaena, which he stated to be greedy of bones beyond all other beasts in his care. . . .

I do not know what more conclusive evidence than this can be added to the facts already enumerated, to show that the hyaenas inhabited this cave, and were the agents by which the teeth and bones of the other animals were there collected; it may be useful therefore to consider, in this part of our inquiry, what are the habits of modern hyaenas, and how far they illustrate the case before us. . . .

They inhabit holes in the earth, and chasms of rocks; are fierce, and of obstinate courage, attacking stronger quadrupeds than themselves, and even repelling lions. . . . The strength of the hyaena's jaw is such, that in attacking a dog, he begins by biting off his leg at a single snap. . . . They live by day in dens, and seek their prey by night, having large prominent eyes, adapted, like those of the rat and mouse, for seeing in the dark. To animals of such a class, our cave at Kirkdale would afford a most convenient habitation; and the circumstances we find developed in it are entirely consistent with the habits above enumerated.

It has been observed, when speaking of the den, that the bones of the hyaenas are as much broken to pieces as those of the animals that formed their prey; and hence we must infer, that the carcases even of the hyaenas themselves, were eaten up by their survivors.

In many of the most highly preserved specimens of teeth and bones there is a curious circumstance, which, before I visited Kirkdale, had convinced me of the existence of the den, viz. a partial polish and wearing away to a considerable depth of one side only. . . . This can only be explained by referring the partial destruction of the solid bone to friction from the continual treading of the hyaenas, and rubbing of their skin on the side that lay uppermost in the bottom of the den. . . .

The extreme abundance of the teeth of water rats has also been alluded to; and though the idea of hyaenas eating rats may appear ridiculous, it is consistent with the omnivorous appetite of modern hyaenas. . . .

With respect to the bear and tiger, the remains of which are extremely rare, . . . it is more probable that the hyaenas found their dead carcases, and dragged them to the den, than that they were ever joint tenants of the same cavern. It is however obvious that they were all at the same time inhabitants of antediluvian Yorkshire.

As ruminating animals form the ordinary food of beasts of prey, it is not surprising that their remains should occur in such abundance in the cave; but it is not so obvious by what means the bones and teeth of the elephant, rhinoceros, and hippopotamus, were conveyed thither [since] the cave is in general of dimensions so contracted (often not exceeding three feet in diameter) that it is impossible that living animals of these species could have found an entrance, or the entire carcase of dead ones been floated into it. . . . [But] though an hyaena would neither have the strength to kill a living elephant or rhinoceros, or to drag home the entire carcase of a dead one, yet he could carry away, piecemeal, or acting conjointly with others, fragments of the most bulky animals that died in the course of nature, and thus introduce

took the form of an anonymous broadsheet, comprising a lithographed cartoon and an accompanying poem (fig. 17; text 10). This was probably distributed widely among the members of the Geological Society in Britain; it certainly reached Cuvier, and probably others too, on the Continent.[13] The poem—or, less charitably, doggerel—celebrates Buckland's work with somewhat ponderous humor and is packed with allusions to his detailed conclusions. The cartoon is a lively rendering of Buckland's verbal reconstruction. The hyenas are feasting on a variety of bones that they have scavenged from the area outside the cave, but also included is a skull of one of their own kind.[14]

However, the picture is no simple reconstruction. Into the cave crawls Buckland himself, candle in hand, illuminating this scene from deep time with the light of science, penetrating the epistemic barrier between the human world and the prehuman, and looking perhaps as surprised to see the hyenas as they are to see him. Buckland's lectures were famous—or, to his more staid colleagues, notorious—for his humorous impersonations of his extinct creatures. So this scene from the almost inconceivable recesses of deep "antediluvial" time is made conceivable by being at the same time a scene of the amiably eccentric Buckland crawling into the cave like one of his hyenas. The scientist becomes an actor *within* the scene that he has reconstructed. The final stanza of Conybeare's poem makes the point clearly. Kirkdale cave is "mystic," in the sense that this feat of penetrating into deep time is like the magic of a fairy tale. Buckland's reconstruction fills in what the bare textual records of the world before the Deluge hardly begin to suggest. Above all, it becomes a keyhole in a locked door, allowing us at least to "spy" a glimpse of a lost world that would otherwise be inaccessible forever.

Conybeare knew perfectly well, of course, that the

hugely thick succession of strata he had just described in his book provided the raw material for the construction of far more than *one* other such keyhole into the deep past. But such poetic license was justified, for Buckland's work presented a crucial example of how any number of keyholes could be made.

However, Conybeare's implicit suggestion was not followed up for several years. Nor was his cartoon itself translated into a more soberly scientific scene from deep time or given more permanent form in a normal scientific publication. When eventually a new scene from deep time was devised, it was, like Conybeare's cartoon, inserted almost stealthily into the scientific world, as another privately distributed broadsheet. The subject was a moment quite different from, and incidentally much older than, either Cuvier's mammals from Montmartre or Buckland's hyena den. What made it possible were empirical circumstances like those of Cuvier's megatherium, namely, a set of fossils of extinct animals that by happy chance were found preserved with their skeletons almost complete.

Early in the century, some relatively complete skeletons of crocodilelike fossils began to turn up in certain Secondary strata, particularly those known by the quarrymen's name of "Lias," at various localities in the south of England. When further specimens showed that the animal had had paddles rather than legs, its zoological affinities became the subject of lively discussion. The name it was given, *Ichthyosaurus* ("fish-lizard"), aptly reflects the uncertainty.

At this point Conybeare, who was then living and working in the region where most of the specimens were being found, began to study the comparative anatomy of these fossils in good Cuvierian style. In 1821—the year that Kirkdale cave was discovered—he read a paper to the Geological Society, not primarily on the ichthyosaur

them to the inmost recesses of his den. . . .

Thus the phenomena of this cave seem referable to a period in which the world was inhabited by land animals, bearing a general resemblance to those now existing, before the last inundation of the earth; . . . the facts developed in this charnel house of the antediluvian forests of Yorkshire, show that there was a long succession of years in which these animals had been the prey of the hyaenas, which like themselves at that time, must have inhabited these regions of the earth. . . .

A farther argument may be drawn from the limited quantity of post-diluvian stalactite, as well as from the undecayed condition of the bones, to show that the time elapsed since the introduction of the diluvial mud has not been one of excessive length.

William Buckland, "An Assemblage of Fossil Teeth and Bones" (1822).

Text 10
Trophonius 'tis said had a den
Into which whoso once dared to enter
Returned to the daylight again
With his wits jostled off their right centre.

But of all the miraculous caves
And all their miraculous stories
Kirby Hole all its brethren outbraves
With Buckland to tell of its glories

BUCKLANDUS IPSE LOQUITUR

Ages long ere this planet was formed,
(I beg pardon) before it was drown'd,
Fierce and fell were the Monsters that swarmed
Roared and rolled in these hollows profound

Their teeth had the temper of steel,
Skulls & dry bones they swallowed with Zest, or
Mammoth tusks they dispatched at a meal
And their guts were like Pappin's digester.

And they munch'd 'em just like Byron's dog
Tartars' skulls that so daintily mumbled

Figure 17. William Conybeare's cartoon of William Buckland entering the den of extinct pre-Deluge hyenas at Kirkdale in Yorkshire in 1821.

but on another, even stranger reptile that he had distinguished among the same set of fossils. This he named *Plesiosaurus* ("almost-lizard"), because he considered it anatomically intermediate between the ichthyosaurs and the crocodiles.[15] But only in 1824, after an almost complete specimen was discovered, did Conybeare publish reconstructions of the skeletons of both reptiles (fig. 18); only then did he feel confident enough to comment—first in private, and then in more staid fashion in print—on their probable mode of life (texts 11, 12). He promptly sent a report of his work to Cuvier in Paris. It arrived just in time to be incorporated in the relevant volume of the new edition of the French naturalist's *Researches on Fossil Bones*, thereby receiving his authoritative endorsement.[16]

In 1829, Buckland repeated his hyena stunt with a detailed study of what he interpreted as the fossilized feces ("coprolites") of these ancient reptiles, thus extending the reconstruction by inferring their dietary habits. He also added to this bizarre fauna by showing that its "bird" bones were really those of the *Pterodactyle* ("wing-fingered"), a strange winged reptile that had already been described from strata of similar age on the Continent.[17] All that was needed for a firmly based pictorial reconstruction of a scene of Liassic life was now available.

The event that precipitated the drawing of such a scene had as its context the deepening economic depression in England in the late 1820s. Many of the Liassic fossils had been coming from Lyme Regis, a small but fashionable seaside resort in Dorset. Some of the finest specimens had been found there by the famous fossil collector Mary Anning (1799–1847). After her father's death left the family almost destitute, she had successfully expanded his part-time business of selling fossils to the aristocracy and gentry who visited Lyme. Developing an outstanding eye for good specimens, she had become

Horns & hoofs were to them glorious prog
Ecce signa—see how they're all jumbled.

I can show you the fragments half gnawed,
Their vomit & dung I have spied;
And here are the bones that they pawed,
And polished in scratching their hide

So unbreeched Caledonians wear out
Each milestone they pass as they go,
So the lip of the pilgrim devout,
Has kissed off St. Peter's great toe.

Some may love potted venison & hare,
Potted char may of some stir the blood;
But no dainty to me is so rare
As "Hyaenas potted in mud."

I know how they fared every day,
Can tell Sunday's from Saturday's dinner;
What rats they devoured can say,*
When the game of the forest grew thinner.

    * For rats and mice & such small deer
     Had been Tom's food for many a year.

Your Elk of the bog was a meat,
That each common hunt might obtain,
But an Elephant's haunch was a treat,
They only could hope now and then.

In scarce winters they sliced up each other,
So gaunt Mariners, struggling with ruin,
Cast lots for each famishing brother,
For particulars Vide Don Juan.

Mystic Cavern, the gloom of thy rock,
Shedding light on each point that was dark,
Tells the hour by Shrewsbury clock
When old Noah went into the Ark.

By the crust of thy Stalactite floor,
The Post-Adamite ages I've reckoned,
Summed their years, days & hours & more,
And I find it comes right to a second.

Mystic Cavern, they chasms sublime,
All the chasms of History supply;

What was done ere the birth-day of time,
Thro' one other such hole I could spy.

William Conybeare, "On the Hyaena's Den at Kirkdale,
near Kirby Moorside in Yorkshire, discovered A.D. 1821"
(1822).

NOTES. *Trophonius*, according to Greek legend, was swallowed up by the earth; those who subsequently consulted him as an oracle, in a cave in Boeotia, emerged in a state of profound melancholy. *Bucklandus ipse loquitur:* Buckland himself speaks. *Pappin's digester:* an early pressure cooker, invented by the French technician Denis Papin around 1679, while he was Robert Boyle's assistant in London. *Prog:* a slang term for food. *Ecce signa:* behold the signs. *Unbreeched Caledonians:* an Englishman's sarcastic comment on kilted Scotsmen migrating south to England. *St. Peter's great toe:* a Protestant's sarcastic comment on Catholic pieties in Rome. *Don Juan:* Byron's satirical poem (1819), in which Juan's dog and then his tutor are eaten by the shipwrecked survivors. *Shrewsbury clock:* a proverbial phrase for accuracy.

Figure 18. William Conybeare's reconstructions of the skeletons of the extinct reptiles *Ichthyosaurus* and *Plesiosaurus* (1824), based on complete skeletons found in the Liassic strata of southern England. Conybeare's drawing was lithographed by George Scharf and printed by Hullmandel.

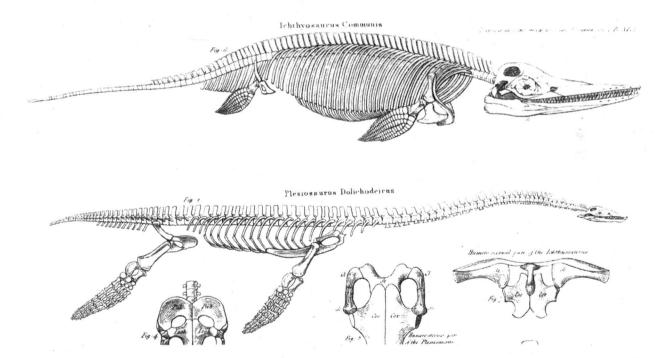

well-known to all the leading geologists and collectors in England.

By 1830, however, there were fewer with cash to spend on such luxuries, her sales fell, and she was in danger of relapsing into acute poverty. One of her customers, who also lived in Lyme and had known her for some years, was the geologist Henry Thomas De la Beche (1796–1855)—an Englishman in spite of his frenchified name—who had in fact been Conybeare's junior co-author. At De la Beche's instigation, some of her gentlemanly friends decided to try to help her financially. De la Beche, who was an accomplished amateur artist, drew for her benefit an imaginative scene, *Duria antiquior* ("An earlier Dorset"), in which the fossils she sold had been alive (fig. 19). The drawing was turned into a lithograph by one of the finest scientific illustrators in London, the theatrical scene-painter George Scharf (1788–1860), who did much of the artistic work for the publications of the Geological Society. Copies of this print were sold to Anning's more wealthy customers, apparently at the substantial price of £2 10s. (about the same price as a large fine-art print such as Martin's *Deluge*), with the proceeds going directly to her. Presumably her benefactors hoped that this lively representation of the extinct animals might also entice new purchasers for her specimens. It seems that the original lithograph soon had to be redrawn, and it was probably printed off in larger numbers and distributed more widely. Buckland's use of an enlarged version of it in his Oxford lectures must have spread its fame still further.[18]

De la Beche's scene is a remarkable achievement. Conybeare's verbal inferences about the ichthyosaurs and plesiosaurs (texts 11, 12) are vividly translated into pictorial terms. The most prominent ichthyosaur is biting the long slender neck of a plesiosaur. Other plesiosaurs are reaching up out of the water to seize a passing

Text 11

[The newly discovered skeleton] made my beast roar almost as loud as Buckland's hyaenas. . . . [He probably] swam on the surface and fished with his long neck, or lurked in shoal water hid among the weeds poking his nose to the surface to breathe and catching all the small fry that came within reach of his long sweep, but he must have kept as much as possible out of reach of *ichthyosauri*, a very junior member of whom with his long powerful jaws would have bit his neck in two without ceremony.

William Conybeare, letter to Henry De la Beche (1824).

Text 12

That it [the *Plesiosaurus*] was aquatic is evident from the form of its paddles; that it was marine is almost equally so, from the remains with which it is associated; that it may have occasionally visited the shore, the resemblance of its extremities to those of the turtle may lead us to conjecture; its motion, however, must have been very awkward on land; its long neck would have impeded its progress through the water; presenting a striking contrast to the organization which so admirably fits the ichthyosaurus to cut through the waves. May it not therefore be concluded . . . that it swam upon or near the surface, arching back its long neck like the swan, and occasionally darting it down at the fish which happened to float within its reach?

It may perhaps have lurked in shoal water along the coast, concealed among the sea-weed, and raising its nostrils to a level with the surface from a considerable depth, may have found a secure retreat from the assaults of dangerous enemies; while the length and flexibility of its neck may have compensated for the want of strength in its jaws and its incapacity for swift motion through the water, by the suddenness and agility of the attack which they enabled it to make on every animal fitted for its prey, which came within its extensive sweep.

William Conybeare, "Skeleton of the Plesiosaurus" (1824).

Figure 19. *Duria antiquior:* Henry De la Beche's cartoon of life in "a more ancient Dorset" (1830). Some animals were identified by numbers on later editions of this lithograph: 1. *Ichthyosaurus vulgaris* (largest, center-right); 2. *I. tenuirostris* (long-jawed, below no. 1); 3. *Plesiosaurus dolichodeirus* (being eaten by no. 1); 4. *Pterodactylus macronyx* (flying overhead); 5. *Dapedium politum* (being eaten by no. 2); 6. *Pentacrinites briareus* (a plantlike crinoid echinoderm, bottom right). Among other animals, note the living ammonite (behind no. 1's back) and ammonite shells (on the sea-floor, bottom left); a squidlike belemnite (being eaten by the lowest ichthyosaur); and the bivalve mollusc *Gryphaea* (on the sea-floor, bottom right).

pterodactyle and to nip the neck of a turtle; other ichthyosaurs are caught in the act of seizing a variety of
fish, whose distinctive scales and bones had been found
by Buckland in their feces. The feces are also shown,
with distinctly pre-Victorian indelicacy, dropping from
several individuals. Among the invertebrates are several
squid, reconstructed from the fossil belemnites common
in the Lias; one is shown being swallowed by an ichthyosaur. In the right foreground is a clump of crinoids
(stalked echinoderms distantly related to starfish), of
which finely preserved specimens were also found at
Lyme Regis. Out on the dry land, there is a somewhat
perfunctory attempt to depict the vegetation as appropriately tropical; the plants—palms, tree-ferns, and a
pineapple-shaped cycad—are based on fossils from
strata of about the same age.

A few dead fossils-to-be are lying on the sea bottom,
most notably two coiled ammonite shells; the curious
butterflylike object floating just behind the largest ichthyosaur is an attempt to reconstruct a living ammonite
as a floating creature rather like the living Paper Nautilus. This conjunction of living and dead alludes neatly
to the process of fossilization that links the reality of the
deep past to the survival of its relics in the present.

De la Beche's scene retains something of the traditional tableau style: the later version of the print even had
the main characters keyed by numbers, like Scheuchzer's
scenes of Creation (figs. 3–5), or indeed Martin's scene of
the Deluge (fig. 10). Unlike Scheuchzer's scenes, however, many of the animals are here shown in intensive interaction with each other, eating or being eaten. Yet it is
no dark apocalyptic picture in Martinesque style, still less
a Darwinian scene of "Nature red in tooth and claw"; it is
somehow much too cheerful for either. The tone is more
Neoclassical than Romantic.

The most intriguing innovation in De la Beche's design

can easily be missed, because it is so familiar to modern eyes. The observer is not out on land, but half in the water, close to the surface and seeing the view both underwater and above it (the realism does not, however, extend to the foreshortening effects of refraction underwater). The viewpoint is as much that of the marine animals themselves as of any ordinary human observer. The source of this striking design is unclear. De la Beche conceived it two decades before the invention of the marine aquarium and the start of the Victorian craze for importing samples of aquatic life into the drawing-room, where a fish's-eye view could be experienced through a sheet of plate glass. But he himself was familiar with boating and diving on the Dorset coast, as were at least some of the visitors to Lyme for whom his picture was designed; he may also have seen contemporary pictures of the new diving devices for salvaging underwater wrecks, some of which used a similar half-subaqueous viewpoint.[19]

Duria antiquior is the first true scene from deep time to have received even limited publication. Cuvier had stopped at mere reconstructions of the body outlines of individual animals; Conybeare had merely distributed a cartoon making a joke of Buckland's detailed imagining of an "antediluvial" hyena den. De la Beche's design, by contrast, depicted a whole range of extinct organisms in a landscape with—despite the jokiness—some degree of realism based on sober scientific analysis of the fossil remains. In that respect, its innovative character and historical significance can hardly be overestimated. In terms of style, however, and apart from its unusual half-underwater viewpoint, it belongs squarely in a well-established pictorial tradition, namely that of natural history illustrations. Not surprisingly, the general composition recalls that of many contemporary illustrations of the living world. Like Cuvier's lateral body profiles, this places the origins of the genre of scenes from deep time as firmly within the traditions of natural history as they are within those of biblical (and, more generally, historical) illustrative art.

De la Beche's lithograph became so famous in geological circles that he was able soon afterward to deploy the same animal figures for a quite different purpose. This was another scene, but from the very opposite reaches of time, namely, the unimaginably remote, post-human *future*. The first volume of the ambitiously titled *Principles of Geology* (1830–33), by the Scottish-born Londoner Charles Lyell (1797–1875), had just set a cat among geological pigeons: not so much for its insistence on the total adequacy of present-day geological processes for explaining everything about the past history of the earth; but more for its claim that that history was in the long term cyclic, rather than leading directionally toward the present. Most geologists found this idea utterly implausible, and De la Beche decided to devise a cartoon that, if lithographed and distributed widely, would help laugh it out of court. After several false starts, he designed a scene that would—and did—fulfill that function.[20]

His cartoon entitled "Awful Changes" (fig. 20) shows an audience of ichthyosaurs and plesiosaurs listening with rapt attention to a "Professor Ichthyosaurus" lecturing on the subject of a *human* fossil skull (Lyell had just been appointed the first professor of geology at the new King's College in London). The cartoon was based on a passage in the *Principles* (text 13), where Lyell had incautiously indulged in a piece of purple prose about the imagined *return* of the extinct reptiles—or at least, of creatures very much like them—in the distant future, as the vast cycle of the earth's history returned to where it had been in Liassic time.[21]

"Awful Changes," as an in-joke designed for the Geological Society circle in London, may not have been known

Text 13

We might expect, therefore, in the summer of the "great year," which we are now considering, that there would be a great preponderance of tree-ferns and plants allied to palms and arborescent grasses in the isles of the wide ocean. . . . Then might those genera of animals return, of which the memorials are preserved in the ancient rocks of our continents. The huge iguanodon might reappear in the woods, and the ichthyosaur in the sea, while the pterodactyle might flit again through umbrageous groves of tree-ferns.

Charles Lyell, *Principles of Geology* (1830).

Awful Changes.

Man found only in a fossil state. —— Reappearance of Ichthyosauri.

"A change came o'er the spirit of my dream." Byron.

A Lecture. —— "You will at once perceive," continued Professor Ichthyosaurus," that the skull before us belonged to some of the lower order of animals the teeth are very insignificant the power of the jaws trifling, and altogether it seems wonderful how the creature could have procured food."

Figure 20. Henry De la Beche's cartoon "Awful Changes" (1830), showing Charles Lyell as "Professor Ichthyosaurus." The quotation from Byron's poem "The Dream" (1816)—to which the title of the car- toon also alludes—is the refrain with which "the Dreamer" moves from one fantastic vision to another; the text of the "lecture" parodies the style of functional anatomy that was commonplace at the time.

widely outside. But *Duria antiquior* certainly spread to the Continent. One copy was seen by Georg August Goldfuss (1782–1848), the professor of zoology and mineralogy at Bonn, and one of the most prominent fossil specialists—they were not yet called paleontologists—in all Europe.[22] But Goldfuss had almost certainly had a similar idea already.

In Heidelberg the previous year, at the annual meeting of the German Association of Scientists and Physicians (*Gesellschaft Deutscher Naturforscher und Ärzte*—soon to become a model for the new British Association for the Advancement of Science), Goldfuss had displayed a superb new specimen of a pterodactyle. This had been found at Solnhofen in Bavaria, where fine-grained limestone was quarried as a building stone, but also for use in the booming new technique of lithography. The Solnhofen stone was already famous for the exceptional preservation of its fossils (three decades later, it yielded the famous reptilelike bird *Archaeopteryx*, which provided persuasive evidence for Darwin's theory of evolution). It was therefore peculiarly appropriate that when Goldfuss wrote up his research on the pterodactyle and other finely preserved reptiles from Solnhofen to have it published by the Academy of Sciences in Bonn, he complemented his text with superb lithographed illustrations of the fossils. These were drawn for him by his colleague Nicholas Christian Hohe (1798–1868), the professor of art at Bonn.[23]

One plate showed Goldfuss's reconstruction of the entire skeleton of the pterodactyle, in the now standard Cuvierian style. Below it, in a literally marginal position, was Hohe's pictorial reconstruction of the flying reptiles, drawn to match Goldfuss's textual inferences (fig. 21; text 14). Some were shown gliding above a calm sea, while one rested against a cliff face by clinging to the rock with the clawlike fingers on its wings. Included in

Text 14

But whichever authority one may be inclined to follow [on the zoological affinities of the pterodactyle], the image of this animal seems in any case more like a painting produced by the unbounded fantasy of a Chinese artist, than a representation of a truly existing product of nature. . . .

The animal could doubtless use its pelvis and rear to adopt a sitting position like a squirrel, and one might even consider this as its usual habit, were it not that its long, downward reaching wings must have obstructed it. If it crept forwards, it would have had the same difficulty as a bat, while the length and weight of the head and neck, and the relative weakness of the rear limbs, would argue against a hopping movement. It is therefore evident that this animal only used its claws to cling to rock surfaces on cliffs, or to trees where they were available, and to climb upwards on steep rock faces. But it could use its wings to fly, and probably hover over the surface of the water, to catch insects and perhaps also aquatic animals. Its wide throat and the weak support of the jaw suggest that the teeth served more to hold the prey than to cut it up. With the help of its long neck, which doubtless was normally held with a backward curl to keep the head in balance, it could however stretch out towards its prey, alter the centre of gravity of the body, and thereby facilitate frequent turns in flight.

August Goldfuss, "Reptiles of the Former World" (1831).

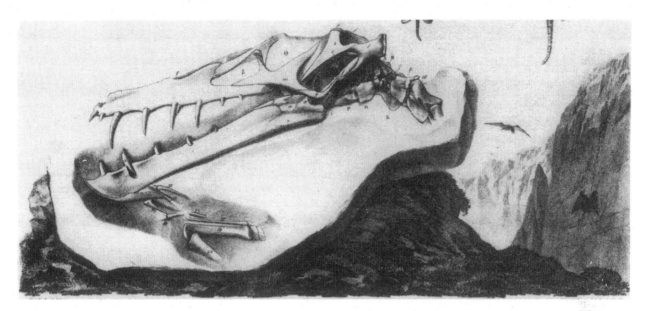

Figure 21. August Goldfuss's reconstruction of pterodactyles from the lithographic stone of Solnhofen (1831), drawn by Christian Hohe, lithographed by the firm Henry and Cohen in Bonn, and published in a corner of a large folding plate showing his reconstruction of the whole skeleton.

this modest representation was part of the specimen on which the scene was based. The skull, embedded in its block of stone, was grossly enlarged to look like a huge boulder perched on the cliff top in the reconstructed scene. Or, to interpret it conversely, the specimen was shown at the same scale as the other illustrations of the skeleton but set in a scene as minute as the past was distant. Either way, the fossil evidence was incorporated, in effect, into the imaginative evocation of its own past, somewhat as Conybeare had incorporated Buckland into a scene of *his* own past.

Buckland had already sent Goldfuss an offprint of his paper on the pterodactyles of Lyme Regis, so it is not surprising that a copy of *Duria antiquior* also found its way to Bonn. De la Beche's ambitious reconstruction evidently spurred Goldfuss to attempt his own scene of the whole range of life at that remote period, into which he could fit his reconstruction of the pterodactyle. The respectable scientific journal that was publishing his paper might not have accepted a whole large plate devoted to a

51

conjectural scene. Fortunately, however, Goldfuss had full control of an outlet of his own. For several years past, he had been publishing by subscription an ambitious illustrated monograph, *Fossils of Germany* (*Petrefacta Germaniae*, 1826–44).[24] As was usual with major works in natural history, it appeared at intervals in successive parts; each consisted of taxonomic descriptions of certain genera and species, with large lithographed plates illustrating the best fossil specimens in Goldfuss's collection and that of his friend Count Georg zu Münster (1776–1844), a wealthy amateur geologist.

When De la Beche's lithograph reached him, Goldfuss was preparing the third installment (1831) of his massive work; like the preceding two, it dealt only with fossil corals, echinoderms, and worms. Undeterred, however, by any qualms about its relevance, Goldfuss at once designed—or got Hohe to design—a scene of life at the time of the *Jura Formation*, the strata that included the Solnhofen stone, and the Continental equivalent of the Liassic strata in England. The scene is unmistakably modeled on *Duria antiquior* and, like it, dominated by the reptiles (fig. 22).[25] To encourage the subscribers to his work, he added this more alluring plate to the prosaic ones of fossil specimens; when the whole first volume was completed later, the scene was bound in as a frontispiece. But it had no obvious pertinence to the volume—except that a few of Goldfuss's invertebrate animals figure subordinately in it—and the text did not refer to it at all.[26]

Goldfuss's scene, as just mentioned, is populated largely by rather inaccurately drawn ichthyosaurs and plesiosaurs, with pterodactyles overhead. The selection of invertebrate life featured in the foreground is much larger than in De la Beche's picture, but most of them are shown simply as shells on the sea bottom, looking much like the fossils in Goldfuss's ordinary illustrations. Only a few are reconstructed in a lifelike manner, and most of

Figure 22. August Goldfuss's "Jura Formation" (1831): a reconstruction of life at the time of what were later termed the "Jurassic" strata, published with his *Fossils of Germany* (1826–44.)

these are borrowed from De la Beche. Above all, Gold-fuss's rendering fails to utilize De la Beche's imaginative underwater viewpoint: in the German design, the submarine life is seen with implausible clarity through the rippled surface of the water.

However, there are several new features. One of the pterodactyles is shown clinging batlike to a cliff, as in Goldfuss's own smaller scene; and the flora is enlarged by the addition of a patch of horsetails. Likewise the fauna is enlarged by the addition of some terrestrial (or semi-terrestrial) forms. The most prominent of these is the *Megalosaurus*, a huge reptile that Buckland had described (but not restored) a few years earlier, on the basis of a few bones found in the Stonesfield "slate" near Oxford.[27]

Most of Goldfuss's design may have been unoriginal, but his picture must have spread the *idea* of such scenes more widely among geologists, beyond the circles that would have seen De la Beche's privately distributed lithograph. Goldfuss's monograph was already an indispensable reference work for anyone concerned with invertebrate fossils from the Secondary formations; and it was subscribed to, or at least consulted by, geologists all over Europe. Apart from Goldfuss's own modest reconstruction of pterodactyles, this was the first true landscape scene from deep time to be published in a straightforward scientific work.

News of Goldfuss's adaptation of *Duria antiquior* soon reached England. It prompted Buckland to urge De la Beche to add some more scenes without delay, before the Germans stole all the best ideas (text 15). This suggestion was of great importance, because for the first time a *sequence* of scenes from several moments in deep time was envisaged: in effect, *several* keyholes into the past. Buckland listed his suggested sequence, not in true chronological order, like a classic sequence of historical

Text 15

I have just received from Hoeninghaus thro' Lonsdale (who may also probably have the same for you) a German parody of your Duria Antiquior which has much merit tho' the notion is entirely borrowed from your original & now that Germans have taken to this sort of thing, I am very anxious that you should not be forestalled by them & write to beg you to put on the stocks in your best style 2 or 3 more restorations of scenes in the ancient world.

I. The Period immediately preceding the formation of Diluvium—a land piece—with only rivers plains & mountains—as in Palestrina Pavement—exhibiting the gamboled Battles of Elephants & Rhinoceros & Mastodons—Hippopotami jumping into the Rivers—Megatherium sitting on his Haunches with one fore Paw against the trunk of a tree and the other reaching down an enormous Branch—Horse Ox and Elk scampering before a Pack of Wolves and falling headlong into fissures—Hyaenas in their Den or dragging into its Mouth their Prey—Tigers crouching to spring on Deer.

II. A lake scene from the F[resh] water Period. Ponds full of Palaeotherium Anoplotherium and Chaeropotamus and all the Paris Pigs of those Days. Dogs and Sarigues at Montmartre. Birds and Reptiles, snakes and water Rats in Auvergne Tortoises Beavers Crocodiles—Volcanoes in the Distance.

III. A sea scene—Sea with tropical Islands of Carboniferous and Transition Periods. Land very short of animals, but glorious hypertropical Vegetation—Sternberg's Lepidodendron, stems of gigantic cactus. Sea full of Tropical islands—covered with Coal Plants—under water, encrinites, Corals, chain Corals—Orthoceratites and Nautili at Surface—Spirifer, Producta—Trilobites—and a few fish. The Trilobites wd caricature well.

I hope you will get out one or two of these proposed scenes forthwith.

William Buckland, letter to Henry De la Beche (1831).

NOTES. Friedrich Wilhelm *Hoeninghaus* (1770–1854) was a fossil collector in Bonn and a friend of Goldfuss's. William *Lonsdale* (1794–1871) was the paid administrator of the Geological Society of London, which had just received a copy of Goldfuss's picture. *Diluvium:* Buckland's term for the enigmatic deposits that he (and many other geologists) attributed to the "geological Deluge." *Palestrina Pavement:* the Hellenistic "Barberini mosaic" in the Italian town of Palestrina (the birthplace of the sixteenth-century composer of that name), featuring the human and animal inhabitants of the Nile valley. *Sarigue:* the French word for an opossum, as used by Cuvier in his work on these fossils from *Montmartre* in the suburbs of Paris. *Auvergne:* the region in central France where freshwater sediments with Tertiary fossils were closely associated with contemporary volcanic rocks. A monograph (1820–38) by Count Kaspar *Sternberg* (1761–1838) was the standard work on the plant fossils of the Coal or Carboniferous strata, among them the striking tree-sized trunks named *Lepidodendron*. *Encrinites*, etc.: a fairly standard list of invertebrate animal fossils from the *Transition* (pre-Carboniferous) formations, soon afterward termed "Silurian."

Figure 23. Henry De la Beche's vignette of life at the time of the Tertiary strata around Paris, based on Cuvier's published reconstructions (fig. 16), and published in the second edition of De la Beche's *Geological Manual* (1832).

or biblical illustrations (figs. 1–7), but in reverse, from youngest to oldest. However, this simply paralleled the conventional layout of most books on stratigraphical geology, which started with the better known, younger strata, and worked toward the older and more obscure: Conybeare's *Outlines* (1822), De la Beche's *Manual* (1831) and Lyell's *Principles* (1833) were all arranged in this way.

What Buckland suggested were scenes (i) from the time of his Kirkdale hyena den, just before the apparent Deluge event; (ii) from the time of Cuvier's Tertiary mammals; and finally, back beyond the scene De la Beche had already produced, (iii) from the time of the coal forests and the associated Transition limestones of marine origin (the "Carboniferous limestone" of anglophone geologists).

Rather surprisingly, De la Beche never took up Buckland's suggestions. All he did, in apparent response, was add three small and modest vignettes to the new (second) edition of his *Manual* (1832). They look curiously out of place in this valuable but dry-as-dust reference work: other illustrations were limited to severely practical drawings of fossils and geological sections.

The first vignette (fig. 23) did indeed show Cuvier's mammals from Montmartre, but in a manner far more cautious than Buckland's lively proposal. It simply reproduced their outlines from Cuvier's *Researches*, still in Laurillard's stiff profiles (fig. 16), as if De la Beche were unwilling to trespass on Cuvier's authority by trying to make the animals more lifelike. That he could easily have done so is clear from the crocodile they are shown contemplating. The plant life is again somewhat perfunctory, but nonetheless the design does add up to a true landscape scene, albeit a modest one.[28] Apart from a note acknowledging Cuvier as a source, there was no descriptive or explanatory text.

Much later in the book, where De la Beche reached what he termed the "Oolitic Group" (which included the Lias, and was the agreed equivalent of the *Jura Formation*, or *terrain jurassique*, of Continental geologists), he drew a pair of small vignettes of ichthyosaurs and plesiosaurs (figs. 24, 25; text 16). But he sacrificed the underwater realism of *Duria antiquior* by showing the reptiles out of the water, explicitly just to make their forms more visible. The concession suggests how unfamiliar an aquariumlike view may still have been among his potential readers. Again, the background topography and vegetation are decidedly minimal.

Like Cuvier's failure to publish his own best reconstructions, De la Beche's decision not to reproduce *Duria antiquior* or any adaptation of it in his ordinary publications should probably be attributed to an underlying hesitation about the scientific propriety of such pictorial scenes from deep time. De la Beche's career was as precarious as Cuvier's had been, since he had lost his earlier wealth and could no longer be an independent gentlemanly geologist.[29] It was therefore important for his future career prospects that his publications should be accepted as impeccably reliable and authoritative. Designing a scene in semi-humorous style and distributing it among his friends was one thing; publishing an intrinsically conjectural representation in a serious scientific work was quite another.

Text 16

It does not seem unphilosophical to infer that the bays, creeks, estuaries, rivers, and dry land, were tenanted by animals, each fitted to the situations where it could feed, breed, and defend itself from the attacks of its enemies. That strange reptile the Ichthyosaurus . . . may, from its form, have braved the waves of the sea, dashing through them as the Porpess now does; but the Plesiosaurus . . . would be better suited to have fished in shallow creeks and bays, defended from heavy breakers. The Crocodiles were probably, as their congeners of the present day are, lovers of rivers and estuaries, and like them destructive and voracious. Of the various reptiles of this period, the Ichthyosaurus . . . seems to have been best suited to rule in the waters, its powerful and capacious jaws being an overmatch for those of the Crocodiles and Plesiosauri. Thanks to Professor Buckland, we are now acquainted with some of the food upon which these creatures lived: for their fossil faeces, named *Coprolites*, having afforded evidence, not only that they devoured fish, but each other; the smaller becoming the prey of the larger, as is abundantly testified by the undigested remains of vertebrae and other bones contained in the coprolites. Amid such voracity, it seems wonderful that so many escaped to be embedded in rocks, and after the lapse of ages on ages to tell the tale of their existence as former inhabitants of our planet.

Henry De la Beche, *Geological Manual*, second edition (1832).

The establishment of the reality of extinction, notably by Cuvier, provided for the first time the raw material for composing illustrations that would depict scenes significantly, and interestingly, *different* from scenes of even the most exotic parts of the present world. Cuvier himself never went beyond one purely textual scene, and never

Figure 24. Henry De la Beche's vignette of living *Ichthyosaurus,* published in the second edition of his *Geological Manual* (1832). "It is attempted in the annexed wood-cut, to convey an idea of the probable form of *I. communis,* and of the head of *I. tenuirostris.* The former is represented on dry land, where it probably never reposed, for the purpose of exhibiting its form."

published his best reconstructions even of individual animal bodies. His English follower Buckland likewise painted a vivid verbal portrait of an "antediluvial" hyena den, but it was translated into visual terms only as a cartoon, and not by himself. De la Beche later composed a pictorial scene of yet another moment in deep time, which was also based on sober scientific inferences, but it too was cartoonlike, and he never published it in a regular scientific format. Even the scenes that Goldfuss inserted in his scientific works, inspired in part by De la Beche's design, were decidedly marginal to those works.

This pervasive marginality suggests that the act of reconstructing an imagined scene from the deep past, however firmly founded on scientific inferences, was initially regarded as unacceptably conjectural. It was unlikely to improve its author's scientific reputation or bolster the authority of his conclusions, unless a jocular style made it clear that it was not to be taken too seriously. However, that stylistic jocularity can also be regarded as a way in which the inhibiting constraints of the antitheoretical scientific climate of the time could simply be bypassed, so

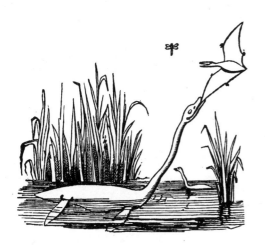

Figure 25. Henry De la Beche's vignette of living *Plesiosaurus,* published in the second edition of his *Geological Manual* (1832). "The animal is represented in the act of catching a *Pterodactyle.* It is figured as swimming high above the water for the purpose of showing its general form. It more probably swam beneath the surface, in the manner of crocodiles, which would enable it the better to support its great length of neck."

that legitimate inferences about the deep past could be communicated effectively to other scientists.

In artistic terms, the style of these earliest scenes from deep time is derived, unsurprisingly, from the well-established tradition of natural history illustrations. That they shared the rather static, tableaulike appearance of some earlier biblical illustrations is simply a result of their common dependence on the same artistic tradition. The most significant feature of biblical illustrations, namely, their arrangement in a temporal sequence, had yet to show itself within the nascent genre of prehistoric scenes. Although this chapter has illustrated scenes, or at least the beginnings of scenes, from three distinct moments in earth history, only Buckland suggested the possibility of arranging them as a sequence; and even he envisaged it as a retrospective sequence, which would have obscured any sense of *geohistorical* development. In any event, he made his suggestion only in a private letter.

That retrospective sequence suggests that Buckland, and probably others too, were fully aware that the construction of any scene from deep time involved an act of imaginative epistemic *penetration* from the known present into the unknown past. Certainly that was what Conybeare made explicit in his cartoon of Buckland crawling into his own antediluvial cave. That cartoon's magical or fairy-tale quality highlighted the problematic character of the act of penetration, as well as the sense of wonder that its achievement evoked. As Conybeare's doggerel suggested, it was an act of "spying," as it were, through a keyhole, into a prehuman past that was otherwise inaccessible to human experience.

3

Monsters of the Ancient World

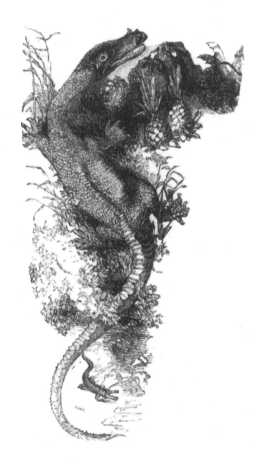

The earliest scenes from deep time, which were reproduced and analyzed in chapter 2, were on the fringes of accepted scientific practice. But what they were marginal to was the practice of *scientists* (the word was only coined in English at just this time). Cuvier's and Goldfuss's weighty monographs would not have been consulted by any but active practitioners of geology and zoology, though few such people were "professionals" in the narrow sense of earning their living by that activity. Likewise, Conybeare's and De la Beche's cartoons were not likely to have been seen by many people outside the circles of gentlemanly "men of science" (and their wives) in which they themselves moved. However, some of those scientific practitioners had no reason to share De la Beche's apparent inhibitions about the scientific propriety of publishing such scenes. On the contrary, authors who were concerned to reach a wider public were quick to see their potential for capturing the imagination of the general reader and, not least, of the young. Within a few years of De la Beche's and Goldfuss's scenes of Jurassic life, adaptations of them had reached a far wider public.

One of the earliest examples adorned the front page

of the *Penny Magazine* only three years after De la Beche's print of "a more ancient Dorset" was first distributed. The high-minded Society for the Diffusion of Useful Knowledge intended this magazine, as its name implies, to place reliable informative material within the reach of the rapidly enlarging literate public in Britain. In this it was spectacularly successful, as its weekly circulation rapidly approached the previously unthinkable figure of 100,000 copies, thanks to the use of the new steam-powered presses and cheap paper. Its pages of dense type were relieved by the generous use of wood engravings, a medium much less expensive, if less subtle, than traditional copper engraving or the newer lithography.[1]

Among the many "men of science" who willingly contributed to the *Penny Magazine* was the young geologist John Phillips (1800–1874), who had learned his science by informal apprenticeship to his uncle William Smith (1769–1832), the self-taught pioneer of the use of fossils in stratigraphical geology. Phillips, who had just become Lyell's successor as part-time professor of geology at King's College in London, rounded off a sequence of short articles on fossils by presenting the magazine's readers with the view "Organic Remains Restored" (fig. 26). The "eminent geologists" on whose work the scene was based were not identified, but the engraving is clearly derived from both De la Beche's *Duria antiquior* (fig. 19) and Goldfuss's "Jura Formation" (fig. 22). For the first time, this moment from the deep past was made vividly real to the general public in Britain.

The following year, a similar scene appeared on the Continent in the *Illustrated Dictionary of Natural History* (*Dictionnaire pittoresque d'histoire naturelle*, 1834–39).[2] The Parisian publishers of this ambitious compilation intended it to gain a share of a lucrative market by embellishing the text with a large number of engraved illustra-

Figure 26. "Organic Remains Restored": a wood engraving to illustrate one of John Phillips's anonymous articles on geology in the popular *Penny Magazine* (1833). "In order to give the reader a clearer idea of the animals and vegetables which characterise the lias and oolitic series of the secondary strata . . . we give a representation of the principal species at present known as restored by some of the most eminent geologists . . .
I. PLANTS. 1. Ferns;
2. Zamia (Cycadeae); 3. Arbor Vitae; 4. Dracaena; 5. Araucaria pine;
6. Mare's Tail (Equisetum).
II. ANIMALS. 7. Dragon Fly;
8. Geometric Tortoise; 9. Megalosaurus; 10. Ichthyosaurus;
11. Plesiosaurus; 12. Ammonitis;
13. Echinus; 14. Nautilus; 15.
Cuttle Fish; 16. Encrinitis; 17. Birdlike Bats (Ornithocephali)."

tions brilliantly colored. Like the *Penny Magazine,* the dictionary was sold in frequent installments to subscribers, who had to have their interest sustained from one issue to the next. The editor and his team of contributors therefore needed to find a steady supply of attractive illustrations.

The articles in successive installments moved relentlessly from A toward Z: the dictionary eventually took six years and nine fat quarto volumes to get there. The articles "Animals, fossil" and "Animals, extinct" were published in 1834 in one of the earliest issues. Those entries were assigned to Emile Le Puillon de Boblaye (1792–1843), a military engineer by training but by inclination more of a geologist: in 1834 he was secretary of the Geological Society of France, which had been founded four years earlier in emulation of the London society. To illustrate his articles, Boblaye adopted independently the same tactic as Phillips, getting his artist to base the design on a free adaptation of elements from both De la Beche's and Goldfuss's earlier scenes.

What the resulting picture lacked in refinement and originality was compensated by the impact of its coloring (not, unfortunately, reproducible here). It was the first scene from deep time in which extinct organisms were shown with their colors as well as their shapes restored. Of course there was absolutely no evidence on which to base the colors. But an uncolored engraving would have been in stark contrast to all the other illustrations and might have made some subscribers feel they were not getting their money's worth. Anyway, Boblaye's scene was presented in exactly the same format as the colored pictures of living animals and plants that embellished every issue of the dictionary. As a result, it achieved a possibly unintended effect.

It must surely have conveyed to a wide range of francophone readers the sheer reality of the deep past, just as

Text 17

We have sought to represent [in fig. 27] some of the most remarkable extinct animals, just from the Secondary epoch. Their external forms are deduced from their skeletons. No. 1 is a figure of the ichthyosaur, whose powerful jaws reach a length of eight feet. The plesiosaur, an animal almost as gigantic, whose long snake-like neck rises out of the water, is figured as no. 2. Nos. 3 and 4 indicate pterodactyles, which are flying reptiles. Under no. 5 is represented one of those gigantic crocodiles found fossil at Maastricht. Nos. 6, 7 and 8 are figures of dragonflies, turtles and ammonites from the same geological epoch. Finally, in the foreground are cycads, tree-like horsetails and other plants of the Secondary epoch. . . .

The gigantic reptiles, more or less related to the lizards, appear in the lower stages of the Secondary formations; these are the ichthyosaurs and plesiosaurs, marine animals surpassing in size anything that the tropics can now show us of this kind; these are joined, in the Jurassic and Cretaceous stages, by a crowd of equally strange reptiles, drawings of which, based on their skeletons, seem not so much reality but rather the work of a diseased imagination.

Emile Boblaye, "Fossil Animals," in *Illustrated Dictionary of Natural History* (1834).

Figure 27. "Extinct Animals": a steel engraving (hand colored in the original) by Johann Baptist Pfitzer from a drawing by de Sainson, illustrating the article of that title, by Emile Boblaye, in Felix Guérin's *Illustrated Dictionary of Natural History* (1834).

Phillips's scene did for the anglophone world. But it must have done so even more effectively, because it showed the prehistoric creatures in exactly the same artistic style as ordinary colored illustrations of natural history. A nearby plate, for example, showed a central American landscape with two tropical birds (and natives and a sailing ship in the background), and a view of European woodland with a grass snake (and fashionable Parisians riding in the background).[3] The implicit message of the colored picture "Extinct Animals" was thus that the ichthyosaurs and plesiosaurs were just as real, and in a sense just as much "present" to the natural historian, as any living organisms. At the same time, subscribers to the dictionary would also have imbibed Boblaye's brief but persuasive argument that the evolutionary ideas of Lamarck, so strenuously opposed by Cuvier until his recent death, were becoming steadily more plausible, not least as a result of the latest research on the succession of fossil forms of life.[4]

Another large scene based on De la Beche's work, although published the same year, was significantly different. Among the most assiduous collectors of the Liassic fossil reptiles was Thomas Hawkins (1810–89), a young Englishman who had inherited a fortune and a country house to match. Living in Somerset, near some of the best Liassic fossil localities, he had amassed one of the finest private collections in the country (he was one of Mary Anning's best customers). But apparently he overstretched himself financially, for while still in his early twenties he sold his fossils—which, he boasted, weighed 20 tons and comprised 4,000 square feet of slabs of Lias rock—to the British Museum.

In this connection, Hawkins published a handsome folio monograph, *Memoirs of Ichthyosauri and Plesiosauri* (1834), to record his collection in visual form. Subtitled *Extinct Monsters of the Ancient Earth*, it was a peculiar mixture of the straightforwardly scientific and the bizarrely idiosyncratic. The work was dedicated to Buckland and, for the part on the plesiosaurs, to Conybeare. Privately, Conybeare told Buckland he thought the book "capital fun" and its author "a Geological bore" to beat any of Walter Scott's characters.[5] But publicly, they conceded its scientific value, for the specimens were drawn by Scharf at his usual high standard. Like more than forty other prominent geologists and aristocratic collectors, Buckland and Conybeare paid in advance for subscription copies of the work.

The value of the illustrations was in practice quite separable from the rambling idiosyncrasies of the text. However, Hawkins was no naive literalist in biblical interpretation. He had evidently read widely in contemporary scholarly discussions about the meaning of the Creation story in Genesis, and he seems to have had at least some knowledge of the Hebrew of the original text. He adopted the common view that its "days" referred to the unimaginably long periods of geological time. But his style in describing the depths of time and the creatures that lived in it is undeniably strange (text 18).

Hawkins's monograph has as its frontispiece a lithographed scene of his own design, which was drawn for him by the landscape painter John Samuelson Templeton (d. 1857). It shows the ichthyosaurs and plesiosaurs whose fossil remains he had collected, alive in their original habitat (fig. 28). The animals are unmistakably derived from De la Beche's published vignettes (figs. 24, 25), and perhaps also from *Duria antiquior* (fig. 19). However, in contrast to the crowded animation of De la Beche's designs, here the scene suggests the bleak emptiness of the prehuman world. But the full peculiarity of Hawkins's design is not evident without his accompanying description. What seems to be a sky lit by the sun behind broken clouds is in fact intended to depict "the angry light of supernal fire," in a "sunless and moon-less" world *before* the creation of those heavenly "lights." Hawkins's aim, which was scarcely achieved by Temple-ton's drawing, was thus to portray a physical world as profoundly alien as the "monsters" that had lived in it. In place of the cheerful jollity of De la Beche's and Gold-fuss's scenes, and of Phillips's and Boblaye's popular ad-aptations of them, Hawkins introduced a more somber note into the depiction of the prehuman world, even if that effect was largely confined to his commentary.

When Buckland received his subscription copy of Hawkins's work, he himself was busy writing *Geology and Mineralogy* (1836) for the educated public. Since it had been commissioned under the will of the Earl of Bridgewater, he could look forward to receiving the munificent fee of £1,000 for his work. The series of "Bridgewater Treatises" was intended to bring up to date a traditional argument in natural theology, by pre-senting modern scientific knowledge in terms of its evi-dence for the divine design of the created world. Most of Buckland's book was devoted, however, not to geol-ogy proper, still less to mineralogy, but to what was just beginning to be termed paleontology. He analyzed the anatomy of extinct animals of all geological ages, in order to show that they too, in their own time, had provi-dentially been as well adapted to their intended environ-ments as any living animals. This argument was designed not least to undercut the supposedly subversive notion—commonly believed to emanate from France, and in fact well represented by Boblaye (text 17)—that the living world had *evolved* by purely natural processes from less well adapted ancestors.

Buckland's massive volume of text was accompanied by a separate one full of fine illustrations. The very first of these was a long fold-out geological section, beau-tifully hand-colored, to give the general reader an im-

Text 18

Our creation is not *a principio,* but from the fourth day or generation of Time, when the lights in the firmament were made "to give light upon the earth." The antecedent history of the planet, as written by Moses, demonstrable by the soundest physics, un-shrouds but the gaunt skeleton of the pre-Adamite epoch, to the clearer comprehension of which noth-ing can so well serve as the accumulation of Fossil Organic Remains. . . .

The globe, sweltering with the intense heat that its primitive revolution in space generated, was a fit-ting habitation for the cold-blooded reptiles, whose day and generation—hid in the AGE before *ages*—may not be computed by us finite [men]. Ichthyo-sauri, which delighted in the depths of the deep sea; ferns and banana-like trees that flourished in the slimy marsh or fringed the sunny lagoon and estu-ary where preyed the wondrous Plesiosauri. The ptero-dactyle too, that paradox which, uniting some of the most specific distinctions of the saurian head with a bird-and-bat-like conformation of body and extremities, has given rise to vagaries of thought as uncertain as the sombre twilight of the ungarnished and desolate world which echoed to the flapping of its leathern wings. They have ceased from off the face of the earth: inexorable time long since ex-tinguished the last of their race and all that survives of these once-grim and omnipotent aborigines are a few crushed bones as unsightly as they are rare. . . .

Theirs was the pre-Adamite—the just emerged from chaos—planet, through periods known only to God-Almighty: theirs the eltrich-world uninhabitate, sunless and moonless, and seared in the angry light of supernal fire.

Thomas Hawkins, *Memoirs of Ichthyosauri and Plesiosauri, Extinct Monsters of the Ancient Earth* (1834).

NOTES. *A principio:* from the beginning (of time). *Fourth day:* of Creation, in the Genesis narrative. *Eltrich:* weird, unnatural, hideous.

Figure 28. "Extinct Monsters of the Ancient Earth": the frontispiece designed by Thomas Hawkins and drawn by John Samuelson Templeton for Hawkins's *Memoirs of Ichthyosauri and Plesiosauri* (1834).

pression of the vast pile of strata from which the *history* of the earth could be inferred. To underline this geohistorical message, a series of vignettes of the successive faunas and floras decorated the space above the section. Yet these took the form of groups of fossils, *not* of scenes from deep time.[6]

In the other sixty-eight plates, Buckland included only one tiny scene (fig. 29), clearly borrowed from Goldfuss's first small reconstruction (fig. 21). It was intended primarily to illustrate his—and Goldfuss's—interpretation of the way the pterodactyles had used their forelimbs for perching as well as flying. Once again, what is most striking is the virtual *absence* of scenes from deep time in a popular work that might seem to have been an ideal vehicle for them. Buckland failed to act on his own earlier suggestion to De la Beche (text 15) and did not give the reading public the vivid *visual* impressions of deep time for which he certainly had the materials. Perhaps he too felt that their conjectural character might detract from the scientific authority of his work. Yet, as demonstrated in his much earlier description of the prediluvial hyenas' den (text 9), he felt no such inhibitions about publishing *verbal* reconstructions. Indeed, his commentary (text 19) on the little scene of the pterodactyles, with its quotation from Milton's *Paradise Lost,* evokes an "infant world" not far removed from the monster-ridden realm of Hawkins's imagination.

Meanwhile, a fossil discovery in Germany had just provided the occasion to enlarge the still limited repertoire of prehistoric scenes. Johann Jakob Kaup (1803–73), a curator at the museum maintained by the Grand Duke of Hesse at Darmstadt, had for several years been studying the fossil mammals from the Tertiary deposits of that part of the Rhine basin. Perhaps the most spectacular was the elephantlike *Dinotherium* ("terrible beast"), as Kaup had named it: the largest land mammal then

Text 19
In external form, these animals [pterodactyles] somewhat resemble our modern Bats and Vampires: most of them had the nose elongated, like the snout of a Crocodile, and armed with conical teeth. Their eyes were of enormous size, apparently enabling them to fly by night. From their wings projected fingers, terminated by long hooks, like the curved claw on the thumb of the Bat. These must have formed a powerful paw, wherewith the animal was enabled to creep or climb, or suspend itself from trees. It is probable also that the Pterodactyles had the power of swimming, which is so common in reptiles, and which is now possessed by the Pteropus Pselaphon, or Vampire Bat of the island of Bonin. Thus, like Milton's fiend, all qualified for all services and all elements, the creature was a fit companion for the kindred reptiles that swarmed in the seas, or crawled on the shores of a turbulent planet.

> "The Fiend,
> O'er bog, or steep, through strait, rough, dense, or
>   rare,
> With head, hands, wings, or feet, pursues his way,
> And swims, or sinks, or wades, or creeps, or flies."

With flocks of such-like creatures flying in the air, and shoals of no less monstrous Ichthyosauri and Plesiosauri swarming in the ocean, and gigantic Crocodiles, and Tortoises crawling on the shores of the primaeval lakes and rivers, air, sea, and land must have been strangely tenanted in these early periods of our infant world.

William Buckland, "Bridgewater Treatise," *Geology and Mineralogy* (1836).

NOTES. *Bonin:* Pacific islands of Ogasawara (north of Iwo Jima), then in British possession. *Milton's fiend:* quoted from "Paradise Lost."

Figure 29. "Imaginary restoration of Pterodactyles, with a contemporary Libellula, and Cycadites": the only reconstructed scene in Buckland's "Bridgewater Treatise," *Geology and Mineralogy* (1836). Apart from the pterodactyles using their claws to cling to the cliff, the organisms mentioned are the cycad plant in the foreground, and the giant dragonfly perched on one of its fronds.

known, with curious tusks curving downward from its lower jaw. When he had exhibited its jaw in Berlin, at the 1828 meeting of the German Association of Scientists and Physicians, he had been urged to publish his work. He had duly started a finely illustrated monograph on this and other fossil mammals in the Darmstadt museum, which was published in French to ensure maximum international attention.[7]

In 1836, however, the exciting discovery of the "colossal skull" of the dinotherium—previously unknown—was an event important enough to merit a separate monograph (in German, but published in French the following year).[8] Several fine folio plates of the new specimen were accompanied by an explanatory text by Kaup; there was also an essay on its geological context by its actual discoverer, August Wilhelm von Klipstein (1801–94), the new professor of mineralogy at the University of Giessen (and therefore a colleague of the famous chemist Justus von Liebig).

Literally outside this conventional scientific report is a matching pair of lithographs which, intentionally or not, encapsulate the character of such research (figs. 30, 31). On the front cover, as part of the title design, is a vignette showing the living dinotherium among its contemporaries in a fine landscape scene (fig. 30). In his text, Kaup agreed with what Buckland had just inferred in his Bridgewater Treatise about the life style of the dinotherium (text 20), and this is clearly the basis for the reconstruction in the vignette. All the other animals in the scene—crocodile, elephant, rhinoceros, horse, lion, deer, and so on—represent other fossils found in the same deposits. In most cases, their remains were much more fragmentary, and they are therefore drawn simply as if they were the living species. Although a lion is chasing some horses in the background, the tone of the scene as a whole is pastoral, centered as it is on the pensive

Figure 30 (*above*). The *Dinotherium giganteum* in its reconstructed habitat in Tertiary time: the vignette, drawn and lithographed by Rudolf Hofmann and Ludwig Becker, on the front cover of the report by August Klipstein and Johann Kaup (1836).

Figure 31 (*right*). The excavation of "the colossal skull of the *Dinotherium giganteum*" from Tertiary deposits at Eppelsheim in Hesse: the vignette, drawn and lithographed by Rudolf Hofmann and Ludwig Becker, on the back cover of the report on the discovery by Klipstein and Kaup (1836.)

E & B

dinotherium. Even the volcano erupting in the background, which alludes to the associated volcanic rocks described by Klipstein, poses no immediate threat. The style is simply modeled on that of contemporary topographical drawings: apart from the absence of humanity, this scene from Tertiary time would have reminded Kaup's readers of rural landscape views of the present world. Nothing could be less "monstrous," but nor was the scene in any sense a jokey cartoon: it was above all an *ordinary* world.

On the back cover of the monograph is a second vignette by the same artists. It is a lively scene of the raising of the newly discovered fossil; a huge femur has already been removed from the trench (fig. 31). Kaup—or perhaps Klipstein—supervises the operation, while the hard work is done by laborers, one of whom is, however, taking time off to celebrate the event appropriately. Playfully, one of the clouds sailing overhead is drawn in the shape of the reconstructed animal: a proboscidean Cheshire cat, as it were, observing its own resurrection.[9]

This pair of vignettes neatly symbolizes the act of constructing any scene from deep time. The "before" of the reality of the deep past is matched by the "after" of its fragmentary survival into the present; or conversely, the "before" of careful excavation is matched by the "after" of imaginative reconstruction. That Kaup included a scene from the depths of Tertiary time in an otherwise conventional monograph is a sign of how he aimed to transcend a purely descriptive paleontology; but that the scene was merely a decoration on the cover is equally a sign of the still tentative and marginal character of the act of reconstruction. Likewise, the unframed vignette signaled a tacit break with the windowlike format of classic representations of landscape; implicitly it suggested a contrast between an imaginative recreation of the deep past and the prosaic reality of the observable bones.[10]

Text 20

I shall confine my present remarks to this peculiarity in the position of the tusks [of the Dinotherium], and endeavour to show how far these organs illustrate the habits of the extinct animals in which they are found. It is mechanically impossible that a lower jaw, nearly four feet long, loaded with such heavy tusks at its extremity, could have been otherwise than cumbrous and inconvenient to a quadruped living on dry land. No such disadvantage would have attended this structure in a large animal destined to live in water; and the aquatic habits of the family of Tapirs, to which the Dinotherium was most nearly allied, render it probable that, like them, it was an inhabitant of fresh-water lakes and rivers. To an animal of such habits, the weight of the tusks sustained in water would have been no source of inconvenience; and, if we suppose them to have been employed, as instruments for raking and grubbing up by the roots large aquatic vegetables from the bottom, they would, under such service, combine the mechanical powers of the pick-axe with those of the horse-harrow of modern husbandry. The weight of the head, placed above these downward tusks, would add to their efficiency for the service here supposed, as the power of the harrow is increased by being loaded with weights.

William Buckland, "Bridgewater Treatise," *Geology and Mineralogy* (1836).

Text 21

I will show you a picture of what creatures were once living where the town of Lyme Regis in Dorsetshire, now stands, and tell you something about their habits of living. You may perhaps be ready to think that a great deal of what we profess to know concerning them, is the work of fancy, but I can assure you it is not, and by and by I will endeavour to convince you that I have grounds enough for what I tell you. . . .

The Icthyosaurus [*sic*] was a great tyrant, and used to prey on every creature that came within his reach; this is known by the fossil remains found in the inside of his body. He used at times even to act the cannibal, and eat his own relations, for a large one has been dug out of the cliff at Lyme Regis, with part of a small one in his stomach undigested; he must have been altogether a very unamiable character. But I shall not say any more about him, leaving you to form your own conclusions from what I have related to you, as his family has been so long extinct, and we ought to say nothing but what is good concerning the dead. . . .

The Plesiosaurus could not have been near a match for its neighbour, the Icthyosaurus, in combat, even when the individuals were of the same size; neither would its form adapt it for cutting through the water so quickly. It must, therefore, no doubt, have often fallen a prey to that voracious monster; perhaps, however, it often played him a trick when he was pursuing it, by running on shore out of his reach; or it might mostly have kept in very shallow water amongst the rushes and reeds, every now and then darting its long neck, like a swan, down at the little fish that came near it; or else suddenly reaching aloft into the air, it may have seized upon some unlucky insect, or Pterodactyle . . . and then laid down as quiet under the rushes as if nothing had happened, waiting for its next mouthful. . . .

These creatures [pterodactyles] used principally to feed upon the large dragon flies, beetles, and other insects of which the remains are found,

Kaup's discovery was so important to Buckland that a picture of the complete skull of the dinotherium featured in the only new plate that he added to his Bridgewater Treatise, when its successful sales called for an almost immediate second edition (1837). Yet instead of reproducing Kaup's complete scene to accompany the drawing of the skull, Buckland simply extracted the figure of the reconstructed animal, which he published in isolation from its inferred environment.[11] Once again, he seems to have balked at publishing any full scene from deep time.

What a leading "man of science" like Buckland still felt reluctant about, however, lesser scientists such as Phillips and Boblaye, writing in a more popular vein, could do with impunity. They were soon joined by others, not least by those aiming their work at the young. "Peter Parley" was the pseudonym used by the American writer Samuel Griswold Goodrich (1793–1860), one of the most successful and prolific authors of books in English for children. His *Wonders of Earth Sea and Sky* (1837) was a remarkably enlightened attempt to capture the interest and imagination of young readers.[12] His policy of dealing with just a few interesting topics in depth, rather than attempting a comprehensive but dry-as-dust compilation of everything, was sufficiently unusual at the time to need explicit justification.

In his introduction to the natural history sciences, "Earth," or geology, is given pride of place. It is significant that, after a very brief explanation of how geohistory can be inferred from the succession of strata, Peter Parley launches at once into telling his young readers "what creatures once lived where Dorsetshire now is." The fauna of the English Lias is thus firmly established as the chief exemplar of what geologists can do. But unlike the treatment of geology in other scientific books for children, Parley's clear description is matched by a lively pictorial representation (fig. 32; text 21). The scene is derived from De la Beche's published vignettes (figs. 24, 25); the description, from Conybeare's or De la Beche's verbal reconstructions (texts 12, 16). The text shows an impressive concern to explain the evidence on which the reconstruction is based, rather than presenting it simply as unquestionable fact. The commentary on the unamiable and cannibalistic ichthyosaurs is charmingly anthropomorphic; in this respect, Parley's text suggests that his young readers can learn about the moral order of nature as much from these long-extinct creatures as from those alive around them.[13]

The following chapter of Parley's book presents a similar reconstruction for the Tertiary of Paris (fig. 33; text 22). But here the author has broken through De la Beche's inhibitions (fig. 23) and released Cuvier's mammals from their iconic stiffness (fig. 16). Parley, or rather his anonymous artist, has made them lively and believable: creatures from deep time, but at the same time quite as real as the horses and dogs that his young readers would have had around them every day.

The following year, a scene from yet another prehistoric period made its appearance in what might seem, at first glance, an unexpected venue. This was the volume *Sketches in Prose and Verse* (1838) by George Fleming Richardson (1796–1848). In his native town of Brighton, a fashionable seaside resort in Sussex, Richardson had until recently been the curator of the Mantellian Museum, which housed the famous fossil collection assembled by the surgeon-geologist Gideon Algernon Mantell (1790–1852). When in 1838 Mantell, like Hawkins before him, sold his whole collection to the British Museum, Richardson was part of the package deal and was appointed to a modest post in the natural history department (the forerunner of London's modern Natural History Museum). As his volume shows, Richardson

and some of which are represented in the picture [fig. 32]. There were also living at the same time with these creatures, several kinds of tortoises, and fish in immense varieties. . . .

We are indebted for a good deal of what I have told you about the animals that once lived where Dorsetshire now is, to a lady, Miss Anning, who spends nearly her whole time in collecting fossils out of the cliffs. No one ought to go near Lyme Regis without visiting her collection.

Peter Parley, *Wonders of Earth Sea and Sky* (1837).

Text 22

I shall show you a picture [fig. 33] representing a state of things much more like the present, than the one we looked at before [fig. 32]. It existed at a later period, though still a great many years ago. . . .

The largest of the animals represented in the plate is called the *Paloeotherium* [*sic*]. . . . It was about the size of a small horse, and must have possessed a little trunk, or proboscis, like the modern Tapir. . . . The smallest was not much larger than a little dog, and you may see the figure of one of them in the picture, going down to the water to drink.

The more slender looking animal is the *Anoplotherium* (or unarmed beast). Its size varied from that of a hare, to that of a large dog; it had a very thick tail like that of a kangaroo. Everything about it would lead one to suppose that it was a timid creature, whose swiftness and agility would protect it against stronger animals; not unlike in disposition to the antelope, or the hare of our times.

When these animals were living, the climate must have been very much warmer than it is at present in France, for their bones are found associated with palm trees, and other vegetable remains of hot climates, and the bones of crocodiles, tortoises, and other creatures which only live in warm regions.

Peter Parley, *Wonders of Earth Sea and Sky* (1837).

Figure 32. "Extinct animals that once lived where Dorsetshire now is": a scene from the *Wonders of Earth Sea and Sky* (1837) by Peter Parley [Samuel Goodrich].

Figure 33. "Extinct animals that once lived where Paris now is": another scene from Peter Parley's *Wonders of Earth Sea and Sky* (1837).

had both literary and scientific aspirations, although by origin he was a tradesman with no formal scientific training.[14]

Richardson devoted two substantial essays to a tour of Mantell's museum, as it had been before its removal to London, and it was to this that his frontispiece referred (fig. 34; text 23). The centerpiece of this scene from deep time, as of the museum, was the extinct reptile *Iguanodon* ("iguana-toothed"), which was Mantell's pride and joy, and his most famous discovery and contribution to science. Unlike all the extinct vertebrates that had figured in earlier scenes, however, this one was *not* known from an almost complete specimen; nor had its skeleton been pieced together, like Cuvier's Montmartre mammals, by painstaking anatomical study of abundant scattered bones. The iguanodon was known only from a small number of precious fragments: a femur, a few vertebrae and bits of rib, some teeth, and a curious conical bone that Mantell had inferred to be a kind of horn on the snout of the animal.[15] The teeth were considered to be close in structure to those of the living iguana, a Central American lizard a few feet in length. Since the fossil teeth were far larger, Mantell simply scaled the iguana up into a terrestrial lizard of gigantic size.

This is the monster that occupies the central place in Richardson's scene. As in De la Beche's *Duria antiquior*, the tone is distinctly cheerful; with such a gleam in the iguanodon's eye, its aggressive posture is hard to take seriously. The poem—or doggerel—with which Richardson summarized the contents of the museum comes likewise to a Panglossian conclusion about the providentially blissful happiness of such creatures (text 23).

The iguanodon is surrounded by the usual reptiles, by now perhaps overfamiliar, which may give the impression that the picture is yet another scene of Liassic life, simply amplified by the huge new terrestrial reptile. In

Text 23
The following lines may, perhaps, be offered as a general sketch of the interesting and valuable contents of this unique collection [in the Mantellian Museum]:—

'Tis indeed a world of wonder,
Found within the earth and under,
Facial forms and wild chimeras,
Creatures of primeval eras,
Startling all our ancient notions,
Showing lands of old were oceans;
Showing oceans once were dry,
As the mountains old and high!
Wondrous shapes, and tales terrific'
Told in Nature's hieroglyphic;
Written in her countless volumes,
Graven on her granite columns!
Showing many a strangest mystery,
From her ancient, wondrous history.

. . .

Yet these giant forms tremendous,
Creatures wondrous, wild, stupendous,—
Huge,—that fancy cannot frame them;
Wild,—that language may not name them,
Differing from a world like this,
Each and all were framed for bliss;
Form'd to share, without alloy,
Each its element of joy,
By that Power that rules to bless,
All were made for happiness!

. . . By the simple process of joining, in idea, these separate structures, placing the vertebrae [of the iguanodon] on the table in the centre of the room, appending the ribs on each side, supporting the enormous trunk on such massive thigh-bones and legs . . . , and then clothing the skeleton with all its investiture of integument, muscles, flesh, skin, and scales, we create a monster which the apartment [in the Mantellian Museum] could not contain as to size, and which was elongated to an extent of eighty or a hundred feet—a whale on land—surpassing all existing realities, and embodying the wildest visions of Eastern fable or romance.

George Richardson, "Visit to the Mantellian Museum," in *Sketches in Prose and Verse* (1838).

THE ANCIENT WEALD OF SUSSEX.

Figure 34. "The Ancient Weald of Sussex," with the iguanodon as its centerpiece: the frontispiece, drawn by George Nibbs and lithographed by George Scharf, of George Richardson's *Sketches in Prose and Verse* (1838).

fact, however, most of the iguanodon bones came from the "Wealden" formation, in the densely wooded Weald region inland from Brighton. The ordinary work of geologists had established that these strata were much younger than the Lias, though also much older than the Tertiary: they belonged in what was beginning to be termed the "Cretaceous" group of Secondary formations (named after the slightly younger and highly distinctive Chalk, widespread throughout northwest Europe). Richardson must have been well aware of these differences of geological age, but probably felt it legitimate to portray all the extinct reptiles as contemporaries. For Mantell had already popularized in England the notion of a distinct "Age of Reptiles," based on fossils from *all* the Secondary formations, to match the later "Age of Mammals" based on fossils from the Tertiary strata.[16]

By the time Richardson's *Sketches in Prose and Verse* appeared in print, Mantell himself had already published his *Wonders of Geology* (1838) with another—but strikingly different—version of this scene (fig. 35; texts 24, 25). Mantell's book was based on his popular lectures at Brighton, and it was in fact compiled from notes that Richardson had taken at the time. What made Mantell's scene so different from Richardson's was the involvement of John Martin.

Martin's painting of the Deluge and the prints that gave it wider circulation (fig. 10) had been so popular that in 1834 he exhibited a new but similar painting on the same theme: when it was shown in Paris, Louis-Philippe awarded him a gold medal. By this time, Martin was evidently aware of geological research. Cuvier is said to have visited his studio while he was working on the new painting, and conversely, Martin was among the stream of famous and fashionable visitors to Mantell's museum (text 24). Mantell urgently needed his popular books to earn him handsome royalties, and he saw at

Text 24

Among the host of visitors who have besieged my house to-day was Mr John Martin (and his daughter) the celebrated, most justly celebrated, artist, whose wonderful conceptions are the finest productions of modern art. Mr Martin was deeply interested in the remains of the Iguanodon etc. I wish I could induce him to portray the country of the Iguanodon: no other pencil but his should attempt such a subject.

Gideon Mantell, *Journal* (1834).

Text 25

THE COUNTRY OF THE IGUANODON

That country must have been diversified by hill and dale, by streams and torrents, the tributaries of its mighty river. Arborescent ferns, palms, and yuccas, constituted its groves and forests, delicate ferns and grasses, the vegetable clothing of its soil; and in its marshes, equiseta [horsetails], and plants of a like nature, prevailed. It was peopled by enormous reptiles, among which the colossal Iguanodon and the Megalosaurus were the chief. Crocodiles and turtles, flying reptiles and birds, frequented its fens and rivers, and deposited their eggs on the banks and shoals; and its waters teemed with lizards, fishes, and mollusca. But there is no evidence that Man ever set his foot upon that wondrous soil, or that any of the animals which are his contemporaries found there a habitation: on the contrary, not only is evidence of their existence altogether wanting, but from numberless observations made in every part of the globe, there are conclusive reasons to infer, that Man and the existing races of animals were not created, until

myriads of years after the destruction of the Iguanodon country—a country, which language can but
feebly portray, but which the magic pen of a [John]
Martin, by the aid of geological research, has res-
cued from the oblivion of countless ages, and placed
before us in all its hues of nature, with its appalling
dragon-forms, its forests of palms and tree-ferns,
and all the luxuriant vegetation of a tropical clime.

Gideon Mantell, *Wonders of Geology* (1838).

Figure 35. "The Country of the
Iguanodon": John Martin's mezzo-
tint for the frontispiece of Gideon
Mantell's *Wonders of Geology* (1838).

once the opportunity for publicity that Martin's interest in the extinct monsters opened up. Apparently the artist did a painting entitled "The Country of the Iguanodon," which hung in Mantell's museum, where it was referred to in one of his lectures (text 25). Turned as usual into a mezzotint print, it was this scene that formed the frontispiece of Mantell's latest popular book, acting as a visual summary of "the wonders of geology" (fig. 35).

The peaceful, pastoral tone of so many earlier scenes has been abruptly replaced by the nightmarish "Gothick" melodrama of the Martinesque style. Three huge reptilian monsters are preying ferociously on each other, watched by a smaller winged one. Although evidently inspired by the iguanodon and pterodactyle, the animals are portrayed with scant regard for anatomical accuracy and are derived more from the long artistic tradition represented by innumerable paintings of "Saint George and the Dragon." Also in the foreground, and more realistic in style, are some turtles, a cycad, and ammonite shells; these are presumably included—in line with Martin's concern for accuracy of scale—to indicate the monstrous size of the reptiles. The landscape stretches down toward a huge river, while palm trees suggest the tropical climate. At least on these points, Martin followed Mantell's inferences about the environment in which the Wealden strata had been formed, and in which his beloved iguanodons had lived.

The involvement of Martin in a scene from deep time brought into conjunction the two pictorial traditions that this book has so far described: that of historical, particularly biblical, illustration, based originally on the interpretation of texts (chapter 1); and that of natural history illustration, based in this case on the reconstruction of skeletons and then whole animal bodies from generally fragmentary fossils (chapter 2). Even more important, however, the application of Martin's style to the nascent genre of prehistoric scenes vastly enlarged the imaginative repertoire available to those who designed such scenes. The deep past could now be depicted as idyllic, or nightmarish, or something in between, with little if any constraint from the prosaic evidence of geology itself.

This interpretative flexibility is perfectly illustrated by the contrast between the two scenes that Hawkins commissioned, before and after Martin's entry into the story. Although Hawkins already had a private vision of the deep past that was close in feeling to Martin's, his earlier scene of Liassic life hardly began to embody what he saw in his imagination (fig. 28). As soon as Martin's scene for Mantell was published, however, Hawkins evidently realized that this was the artist who could best express his intentions. (He must surely have been familiar already with Martin's work, but he may have thought the artist would not tackle a scene that could not be based on classical or biblical sources.) Anyway, for a second monograph on his Liassic fossils, with an even more bizarre text than the first, Hawkins commissioned Martin to draw a new frontispiece (fig. 36; text 26).[17]

The scene is even more Martinesque than the one for Mantell but is clearly derived from it. The ferocious monsters are now evidently based on ichthyosaurs and plesiosaurs. But there is as little regard for anatomical accuracy as before: for example, all three reptiles are given webbed feet instead of paddles. One plesiosaur is provided with the forked tongue of a snake, which certainly increases its fearsomeness. The luridly illuminated pterodactyle in the foreground is much enlarged from the previous design, and this only heightens the anatomical shortcomings of its wings and body. It may seem inappropriate to apply such prosaic criticisms to a picture of undeniable imaginative power; but it does serve to underline the contrast between this kind of scene and the careful reconstructions on which scenes in the other pictorial tradition were based.

The contrast between Hawkins's scenes before and after the involvement of Martin is exactly matched in the case of Richardson's works. His *Geology for Beginners* (1842) had as its frontispiece another design by Martin, this time explicitly entitled "The Age of Reptiles" (fig. 37), which is quite different from the scene in the book he had published only four years earlier (fig. 34). It clearly combines elements drawn from Martin's designs for both Mantell and Hawkins, although it lacks the tonal subtleties of the two mezzotints. However, when the following year Richardson published a new edition of his book, he abandoned Martin's design, perhaps because he felt it had been upstaged by the more dramatic scene that Martin had drawn for Mantell. In any case, he reverted to his earlier artist, George Nibbs, who produced yet another version of "The Age of Reptiles" (fig. 38; text 27), with an obvious debt to De la Beche's pre-Martinesque style.

By this time, the use of scenes from deep time in popular books on the "wonders" of geology was well established, and no self-respecting author would omit to include at least one. But even if there were several, they were not necessarily used to demonstrate that "the former world" had been differentiated into a sequence of distinct epochs. For example, the German writer Carl Hartmann included three scenes (not reproduced here) in his *Created Wonders of the Netherworld* (*Die Schöpfungswunder der Unterwelt,* 1841), which were adapted from De la Beche's now standard scene of Liassic life, his rendering of Cuvier's mammals, and Kaup's scene of the dinotherium. But they were merely included at the end of a highly miscellaneous collection of pictorial images of caves, mines, volcanos, and other "remarkable things" (*Merkwürdigkeiten*).[18]

Text 26

Looking back retrospective far over the wintry Ocean, into Pre-adamic Shades, we encounter execrable and dreary things in the abounding Chaos. Through briny clouds incumbent impetuous Monsters gleam phrenitic, livid or green, or swarthy snakes, quadrupedal and deadly. Wide over the desolate Seas warring Dragons innumerable and hideous, enacting Perdition.

If these great Sea-Dragons certainly suckled their impy brood, which these appearances incline us to believe, [John] Martin has barely attained, with all his stupendous Powers, the utter hideousness their own.

Thomas Hawkins, *Book of the Great Sea-Dragons* (1840). NOTES. *Phrenitic:* delirious, affected with brain fever. *These appearances:* fossil evidence that they were viviparous.

Figure 36. "The Sea-Dragons as
They Lived": John Martin's mezzo-
tint for the frontispiece of Thomas
Hawkins's *Book of the Great Sea-
Dragons* (1840).

Another elementary book on geology, published the same year, likewise had several scenes from deep time; but in this case, they formed an explicit sequence. Although the book was English, and its author must have known of Mantell's book, it shows no Martinesque influence at all. However, its design incorporates a different and equally important innovation. Joshua Trimmer (1797–1865) was a geologist who had just returned to London after several years' experience of mining and quarrying in Wales. His *Practical Geology and Mineralogy* (1841) was, as the title indicates, directed toward the applications of those sciences; unlike Hartmann's book, it was not for recreational reading. Nonetheless, Trimmer chose for his frontispiece an illustration of the "wonders" of geology, rather than anything more prosaic. With Mantell's highly successful book as a model and perhaps also a commercial rival, an elementary geology book such as Trimmer's could only benefit by having its own scene from deep time, however irrelevant to its contents.

In fact, however, Trimmer's frontispiece is not one scene but four (fig. 39; text 28). Four geological epochs, explicitly chosen because they have a fuller fossil record than others, are each represented by a scene; and the four scenes are piled on top of one another, with the oldest at the bottom, just as the corresponding strata are piled on one another in the field. In other words, Trimmer took the conventional format of a vertical geological section, showing a column of strata in the order of their natural occurrence, and adapted it to indicate unambiguously a *sequence* of four corresponding scenes from the history of life.

The scenes themselves are fairly crude; but what is striking is that only for one or two of them did Trimmer and his artist, John Whichelo (1784–1865), have any pictorial precedents. The first (and lowest in position), which depicts life in the Coal or Carboniferous epoch, is

Text 27
DESCRIPTION OF THE FRONTISPIECE

This delineation, which was designed, drawn, and engraved on the wood by Mr Nibbs, represents the shores of that ocean by which the strata of the oolite and lias were deposited. Rocks appertaining to these formations constitute the heights and cliffs; and the vegetation consists of those trees and plants the remains of which are discovered in these deposits, including palms, tree-ferns, pandani, and coniferous trees; together with smaller plants, as the ferns, cycadeae, &c., &c.

The reptiles comprise the ichthyosaurus in the act of devouring a fish; the plesiosaurus, which has seized a pterodactyle, or flying reptile, on the wing; together with crocodiles and alligators, which are depicted on the shores. Turtles and tortoises are prowling on the banks, and the waters of this primeval sea are tenanted by corals, shells, crustacea, and fish, appropriate to this peculiar period of the history of nature.

The artist, it is obvious, has equalled the spirit and vigour of the design by the strictly correct and successful execution of its details.

George Richardson, *Geology for Beginners*, second edition (1843).

Text 28
EXPLANATION OF THE FRONTISPIECE

The frontispiece, for the design of which I have to acknowledge my obligations to Mr. Whichelo, represents the condition of the terraqueous surface, as to vegetation and vertebrated animals, during four remarkable geological epochs. In the lower compartment the land is seen clothed with lepidodendra, sigillariae, and other plants of a tropical aspect peculiar to the carboniferous era. The only contemporaneous vertebrated animals at present known are fishes, all distinguished by the continuation of the vertebral column into the upper lobe of a tail divided into two unequal lobes. Some of them were of

Figure 37. "The Age of Reptiles": John Martin's steel engraving for the frontispiece of the first edition of George Richardson's *Geology for Beginners* (1842).

Figure 38. "The Age of Reptiles": George Nibbs's wood engraving for the frontispiece of the second edition of George Richardson's *Geology for Beginners* (1843).

quite novel. The second, of life in the Liassic or "Oolite" epoch, is based as usual on De la Beche's work. The third, showing Cuvier's mammals from the early Tertiary epoch, is probably also based on De la Beche; but the animals, like Peter Parley's, are now shown in a quite realistic full landscape. The fourth and youngest scene, of Cuvier's mammals from the later Tertiary, is again quite new.

How widely Trimmer's illustration was noticed is unclear; certainly his visual sense of the *diversity* of deep time is hard to match in other publications at this time. It may be no coincidence that the scenes he chose were precisely those that Buckland had suggested privately to De la Beche a decade earlier (text 15). De la Beche had failed to act on that suggestion, but there may have been continued talk about it at the Geological Society in London, of which Trimmer too was a member.

Goldfuss was likewise unaffected by Martin's impact; he was either unaware of it or, more probably, rejected it outright as grossly inaccurate and unscientific. In 1844 he published a new scene, portraying more authoritatively and in far greater detail the same epoch as the first of Trimmer's set. Perhaps as a result of the recent death of Count Münster, whose great fossil collection he had utilized in his work, Goldfuss decided to bring his *Fossils of Germany* to a close. Although he was now preparing the last part of the third massive volume, the work as a whole was still far from complete: since the first volume, with its incongruous scene of the *Jura Formation* (fig. 22), he had still only dealt with the bivalve and gastropod molluscs. In a final gesture to his subscribers, he closed the work with a two hundredth plate depicting a scene of life at the time of the Coal Measures (fig. 40; text 29).

Goldfuss's magnificently detailed lithograph, like his first modest scene many years earlier (fig. 21), was drawn

great size, exhibiting a higher organization than the majority of the fishes belonging to more recent epochs, and indicating an approach to the saurian structure. From the prevalence of these large sauroid fishes, this may be called the megalichthian age.

The poikilitic era, during which true saurians first make their appearance, is passed over because of the paucity of its organic remains; and the second compartment from the bottom exhibits the flora and fauna of the oolites. The vegetation is still impressed with a tropical character, but constitutes a different group, both as to genera and species, from that of the coal strata.

The ichthyosaur and plesiosaur are seen sporting in the waters, crocodiles basking on the shores, the bat-like pterodactyls flitting through the air, and the huge iguanodon feeding on ferns, clathraria, and cycadeae. This, which may be called the saurian age, terminates in the cretaceous system, of which the lower beds alone contain any of these strange reptiles, while the supracretaceous strata afford genuine crocodiles, approaching to existing types.

The cretaceous era is passed over like the poikilitic because its characteristic fossils are exclusively marine, and the third compartment exhibits the plants and animals of the early tertiary, eocene, or palaeotherian age.

The vegetation still indicates a high temperature in the latitudes of London and Paris, by the prevalence of palms intermixed with coniferae and other exogenous trees, approaching the character of existing species.

Extinct genera of the pachydermatous order of mammals abound; of these the anoplotherium commune, as restored by Cuvier, is seen on the left, and behind him are palaeotherium magnum and palaeotherium minus; near the latter is a land tortoise; and it appears from recent discoveries that monkeys might have been represented gamboling on the boughs, and boas coiled around the trunks of the trees. Birds, of which some traces occur in older

Figure 39. "Condition of the terr-aqueous surface, as to Vegetation and Vertebrated Animals, during four remarkable Geological Epochs": John Whichelo's engraving for the frontispiece of Joshua Trimmer's *Practical Geology* (1841).

87

for him by his artistic colleague Christian Hohe. It shows a profusion of the varied terrestrial vegetation of the Coal Measures, combined incongruously with a selection of fossils from the somewhat older Carboniferous Limestone formation. Of these marine animals, only a few fish seen through the clear water, as well as some corals left stranded by the low tide, are shown in lifelike manner; the rest are just shells on the shore, mere fossils-to-be, following the pictorial tradition that stretches back through Goldfuss's own earlier scene (fig. 22) to Parkinson's and even to Scheuchzer's designs (figs. 9, 8).

For the plants, however, Goldfuss's design is more imaginative. Since the distinctive forms of stems and leaves were almost always found in the Coal strata as separate fossils, and rarely in direct association, it would have been sheer guesswork to show specific leaf forms attached to specific kinds of trunk. So the top edge of the scene cuts off the tree-sized plants below their foliage, which is shown separately as fronds broken off and strewn on the ground. Only in the background, too far away to have to be specific, are the trees shown complete. The plants—though not the animals—are minutely keyed to the explanatory text, as was done earlier, for example, in De la Beche's *Duria antiquior*. The text itself is equally detailed and clearly reflects its intended readership of geologists, botanists, and serious fossil collectors.

A professional evaluation of the scenes that were now available is well expressed in the work of the leading francophone paleontologist François Jules Pictet de la Rive (1809–72), the professor of zoology at Geneva. As he compiled his *Elementary Treatise on Palaeontology* (*Traité élémentaire de paléontologie,* 1844–46), Pictet must have considered whether to include any scenes from deep time, and if so, which. Most of his massive four-volume reference work and textbook is devoted to a systematic review of fossil organisms in biological terms; the sections

strata (the new red sandstone, the wealden, and the chalk), are now abundant.

The volcano in the distance represents the craters of Auvergne, at present dormant, some of which commenced their action towards the close of this era, and afford the first decided evidence of sub-aërial eruptions.

The upper compartment exhibits the elephantoid age. In the foreground are seen the elephant, rhinoceros, and hippopotamus (pachydermata of existing genera, but extinct species), which commenced with the miocene or middle tertiaries, and disappeared from Europe at the erratic block period; in the background are placed the stag, ox, and horse, to intimate the extensive developement of those genera during the pliocene era. The hyaena entering his den indicates the increase of carnivora during this period, and the accumulation of mammalian bones in caverns. The forests consist of oak, fir, birch, poplar, and other trees, closely approaching, if not identical with, indigenous European species. The distant volcano represents the greater part of the eruptions of Central France, of the Rhine, Catalonia, and Hungary, which appear to have taken place during the miocene and pliocene epochs.

Joshua Trimmer, *Practical Geology* (1841).
NOTES. *Megalichthian age:* other geologists' "Carboniferous." *Poikilitic:* commonly used for the New Red Sandstone and its equivalents, comprising Murchison's "Permian" (first named the same year) and Continental geologists' "Trias." *Palaeotherian age:* Lyell's (and other geologists') "Eocene." *Elephantoid age:* Lyell's (and other geologists') "Miocene" and "Pliocene." *Erratic block period:* what Agassiz had just claimed as an "Ice Age."

Text 29
OVERALL VIEW OF THE COAL PERIOD
In this picture [fig. 40] the artist has aimed to assemble, through representatives of their genera,

Figure 40. "Overall View of the
Coal Period": Christian Hohe's final
lithograph (1844) for August Goldfuss's *Fossils of Germany* (1826–44).

on vertebrates include many reconstructions of the bodies of extinct animals, in the manner that Cuvier had pioneered (fig. 14), but not of whole scenes of those animals in their habitats. Scenes from deep time would belong, if anywhere, in the concluding part of the last volume, which reviews more briefly the history of life in stratigraphical terms. Pictet's sources were in general so comprehensive that he must have known of most of the scenes that have been reproduced here, except perhaps some of those in popular works. But he chose to include only two.

Pictet commissioned new versions (not reproduced here), much reduced in size but otherwise accurately redrawn, of De la Beche's *Duria antiquior* (fig. 19) and Goldfuss's new Coal forest scene (fig. 40); his artist did, however, tastefully omit the feces from the former![19] Pictet's judgment was sound: these were the two that were based most thoroughly on fossil evidence. Goldfuss's design thus reached a much wider readership than would ever have seen the original in his massive monograph. De la Beche's, after many years of semi-private circulation, and publication only in Phillips's and Boblaye's modified versions and in Hartmann's popular book, achieved at last the status of full publication in a mainstream scientific work.[20]

Most writers of more popular works, however, continued to be little concerned with the niceties of scientific authenticity in their use of scenes from deep time. In his massively illustrated *Gallery of Nature* (1846), for example, the prolific popularizer Thomas Milner introduced his section on geology with a decorative design that well reflects the general view (fig. 41).[21] The reptile that crawls up the margin of the page combines the ichthyosaur's eye with the iguanodon's horn on its snout, with scant regard for accuracy. More significantly, however, this Secondary reptile and the pineapplelike cycads

the organic creation that lived on earth at the time of the Coal Formation.

In the foreground one sees a coral reef emerging from the sea bottom, in the construction of which *Cyathophyllum, Antophyllum* and *Syringopora* were active. A *Pentremites*, broken off from its stem, is reminiscent of its class at the present day, and mollusc shells lie around on the shore. Among the bivalves one recognises scallops and cockles (*Avicula, Isocardia, Cardium*), but predominantly genera of brachiopods (*Productus, Spirifer, Orthis, Terebratula*). The gastropods are represented by *Dentalia*, whelks, *Nerita*, and *Euomphalus* (*Dentalium, Natica, Buccinum, Melania, Pleurotomaria, Trochus, Euomphalus*); and the cephalopods by *Bellerophon, Orthoceratites, Cyrtoceratites* and *Goniatites*. Between them a trilobite can be seen, indicating the early development of the arthropods, and some obliquely scaled fish at the edge of the water point to the ascent of the animal creation even to the vertebrates.

At the edge of the primaeval forest, the fallen vegetation of which has provided the coal seams of our Coal strata, one sees slender, tubular, scarred trunks with palm-like fronds, as well as forked ones covered with little leaves, a vegetation similar to that of our humid tropical islands. These are quite treelike cryptogamic vascular plants, mostly of a giant size, among which monocotyledons only appear scattered very sparsely, and dicotyledons are missing altogether. The slender, regularly and finely striped trunks are *Calamites*, which reach a height of 10–12 feet and belong to the family of our little horsetails.

Nearby and in front of them one sees two broken-off trunks, thickly covered with triangular scars, each of which is on a vertically rhomboidal leaf-cushion. It was thickly covered with small lancet-shaped leaves, and splits into fork-shaped branches, letting one see beneath it a smaller tree. These *Lepidodendra* belong to the family of *Lycopodiacea*, which with us are only moss-like, and in tropical countries do

Figure 41. "Geology": vignette from Thomas Milner's *Gallery of Nature* (1846).

beside it form the foreground to a scene that further back is unmistakably of Tertiary age, with dinotherium, elephants, and even a monkey. The vignette epitomizes the way that all the periods of earth history were still run together, at least imaginatively, in the public mind.

That general impression makes one exception all the more striking. Just before mid-century, one design appeared that attempted in a highly innovative manner to convey a sense of a living world that had *changed* progressively through time. Although the intention recalls Trimmer's earlier and equally innovative design (fig. 39), the new one seems to have been devised quite independently. It was published in London by James Reynolds, an enterprising and prolific author of popular informative works of all kinds, for which the Victorian public had an apparently insatiable appetite. It was one of a large number of pictorial designs that Reynolds sold singly as broadsheets but also assembled into sets on various subjects.

This scene has as its title the singular term that was in the public mind: it is of *the* "antediluvian world" (fig. 42). But there is no visual or verbal reference to any Deluge: the term simply refers to the world "prior to the creation of man." That popular conception is enlarged and informed, however, by the reference to the *different* epochs involved, which is what the scene is designed to show.

Reynolds's artist, the London mapmaker and illustrator John Emslie (1818–75), drew an ingenious scene that is at once single and multiple. Like the essentially decorative drawing for Milner's popular book (fig. 41), published only three years earlier, it places the creatures of distinct periods within a single scene, but here the mixture is systematic and deliberate. As in Trimmer's design, deep time moves from the bottom toward the top, but the sequence here is not a set of discrete scenes. Emslie's design demands or at least encourages a sense of continuity

not reach more than three feet in height, whereas they then grew into 40-foot trees. A plant that is highly remarkable on account of its strange structure [*Stigmaria*], and that forms a major component of the Coal strata, can also be assigned to this family. From its low, dome-shaped central trunk, which is 3–4 feet in diameter, forked branches radiate horizontally outwards on all sides. They reach a length of 9–15 feet and are covered with rounded leaf scars in quincunx arrangement, which are sometimes carried on rhomboidal warts. They serve as buds of single or forked, linear fleshy leaves.

The remaining trunks are tree-like ferns, which now occur in similar form in humid tropical countries. They were however larger than those now living, in that not infrequently they reached a height of 60 feet and a thickness of 2–5 feet, and bore smaller but much more abundant leaves.

These tree-fern trunks are mostly simple, occasionally forked towards the top, and covered over the whole surface with flat rhomboidal leaf scars, which form alternately constricted longitudinal rows, and appear now broad and rimmed, now close together and angular [*Sigillaria*]. On many trunks the scars are small, single or paired, and on half-cylindrical ribs (*Syringodendron*). In the picture we find the foliage not on the trunks but fallen and strewn on the ground, because they are not found in the Coal Measures in their original relationship, so that one has to give them separate names. The varied division of the nerves in the delicate pinnate leaves, by which one identifies many species, cannot be shown at reduced scale [i.e., in the picture]; and for just this reason the artist could not make distinguishable all those whose names remind one of the well-deserving men who have been active in these forests in the service of science [*Neuropteris, Sphenopteris, Phlebopteris, Pecopteris, Cyclopteris, Gleichites, Anomopteris, Odontopteris, Schizopteris, Hymenophyllites,* and *Glossopteris*].

One of the small and not uncommon plants with

Figure 42. "The Antediluvian
World": a broadsheet drawn and
engraved by John Emslie and published by James Reynolds (1849).

93

about the history of life on earth. By a clever use of perspective, the scene is all foreground (except at the very top). The eye moves across a changing panorama of animals and plants, just as the spectators at London's Panorama and similar commercial shows could sweep their gaze across views of distant and exotic places, and just as the countryside swept past the eyes of travelers on the rapidly expanding network of railway lines.[22] Just as such views might pass along an indented shoreline, so this panorama of earth history runs sometimes past groves of trees and herds of animals, but at other times past beaches with drifted shells and clear pools with fish and swimming reptiles. Reynolds's—or rather, Emslie's—pictorial sources need no special comment. Similarities of animal pose make it easy to detect, or at least suspect, borrowings or adaptations from many of the scenes reproduced in this and the previous chapter of this book, many of which Reynolds is likely to have known.

Like Trimmer's sequence, this one can be read either way: the eye can survey the panorama down the page, as it would read a page of text, penetrating the sequence of epochs backward from the relatively familiar creatures of the Tertiary; or it can read the design from bottom up, forward in time from the earliest period with abundant fossils, namely those from the strata that the London geologist Roderick Murchison (1792–1871) had defined a decade earlier as the "Silurian." Those alternatives are of course parallel to the ways the eye can study the column of strata on which the sequence is based. Reynolds published just such an idealized column on a companion broadsheet entitled "Popular Geology" (1849). His comment there on the significance of fossils, on which of course his scene is based, is predictable in its emphasis on the vast scale of deep time; its conventional theistic reference would have reassured his potential purchasers (text 30).

stellate leaves (*Sphenophyllum*) appears to belong to the *Marsiliacea;* the family of another (*Annularia*) is still doubtful. Palms were found extremely rarely in these forests, and so we find in the picture only a single palm branch, which is named after a friend (*Noeggerathia*). Yet another plant (*Dechenia*) is noticeable nearby, perhaps related to the Euphorbias, which, if it also derives from the oldest period of plant life, can be regarded as native to the Coal forests too.

August Goldfuss, *Fossils of Germany* (1844).

NOTE. *Well-deserving men:* an allusion to those whose personal names are commemorated in the specific names of these fossil plants: among the species listed by Goldfuss are some named after Élie de Beaumont, Brard, Brown, Buckland, Hermann, Hoeninghaus, Humboldt, Mougeot, Phillips, Scheuchzer, Schlotheim, Serle, and Zobel. *Noeggerathia* was named after Goldfuss's friend and colleague Johann Jacob Nöggerath (1788–1877), the professor of mineralogy at Bonn.

Text 30

Fossils are the records showing that our Globe was the seat of animal & vegetable life through a countless series of ages before its occupation by the human species—that successive races have flourished, decayed & altogether vanished, that the surface of the earth has undergone numerous changes each separated by an interval of vast duration, that deep seas have formerly occupied the place of our present continents, and that a degree of heat equal to, or greater than that of the equator, must have prevailed in our northern latitudes. The results of Geological investigation are in perfect harmony with the statements of Revelation, and tend to exalt our ideas of that Almighty Being who formed the earth, and all things therein.

James Reynolds, "Popular Geology" (1849).

How widely Reynolds's ingenious design was distributed is unclear. The present rarity of the print may just reflect the ephemeral character of his work in general; but on the other hand, the lack of later imitations of his design suggests that it may never have become widely known.

De la Beche's scene of Jurassic (Liassic) life, in Goldfuss's extended version, was quickly adapted by Phillips and Boblaye to a much wider range of readers, both in Britain and on the Continent, while the pseudonymous Peter Parley began to make it familiar to children too. In the public imagination, this single scene soon came to symbolize the *whole* of the deep past. However, the repertoire began to be broadened by the addition of Kaup's scene of Tertiary mammals, Richardson's and Mantell's scenes of Cretaceous (Wealden) reptiles, and Goldfuss's scene of a Carboniferous (Coal Measures) forest. But all these were still presented as single scenes, rather than as a temporal sequence. Only in Trimmer's and Reynolds's modest but innovative designs, in which successive scenes were either piled on top of one another, like the formations from which they were derived, or woven together into a continuous panorama, was there the germ of a pictorial sense of the temporal development of the prehuman world.

During the first decade after De la Beche's rendering of a "more ancient Dorset," the genre began to reach a much wider public and to include other periods of earth history in its repertoire. It also greatly broadened its imaginative scope, particularly as a result of Mantell's recruitment of John Martin. The deep past became, at least for those who wished it so, a profoundly alien realm inhabited by creatures that were depicted as monsters. This

Romantic or "Gothick" interpretation, previously confined to verbal expression, was now dramatically translated into pictorial terms. Mantell, for one, evidently believed this alliance between geology and the Romantic imagination would promote the popularity of the science (and of course the sales of his books). While more prosaic styles helped convey the sheer reality of the deep past that geology revealed, Martin's scenes certainly served to promote a sense of the otherness of a world without human beings.

From this point of view, the geologists' careful distinctions between different periods of prehuman history hardly mattered. For the wider public, all those periods constituted *collectively* "the ancient world," *die Urwelt*, or *l'ancien monde*. Those phrases, all in the singular, expressed the sense that it was unified by its complete contrast to the present, and above all by the absence of human beings. The discovery and reconstruction of "the ancient world" was now well established as a major part of the Victorian "wonder of science." Pictorial scenes from deep time were what began to make the reality and the "romance" of the prehuman world both widely known and vividly apparent.

4

A First Sequence of Scenes

The general reading public may have thought of deep time in terms of a single undifferentiated otherness; the geologists, on the other hand, were well aware that each major period or epoch had had its own distinctive character, which at least in principle could be represented in pictorial form. That more differentiated vision had so far been embodied only in Trimmer's and Reynolds's illustrations (figs. 39, 42). However much credit they—and their artists—may deserve for the originality of these designs, their impact was probably limited. The work that first made the scientific conception accessible to the public, in a far more impressive and persuasive way, is for just that reason so important that it deserves a chapter to itself.

Sometime around the mid-1840s, the Austrian botanist Franz Xaver Unger (1800–1870), who was then professor of botany at Graz, gave a lecture course on geology. Although quite active in research on both botany and geology proper, it was his study of fossil botany that had given him a growing reputation in scientific circles: he was for example, publishing in installments the handsomely illustrated *Flora of the Former World* (*Chloris protogaea*, 1841–47), a work quite closely analogous to Goldfuss's *[Animal] Fossils of Germany*.[1]

Among the previously published scenes from deep time, which Unger might have borrowed to illustrate his lectures, he is sure to have known the only one that had focused on plant life rather than animals, namely, the Coal forest scene with which Goldfuss had recently rounded off his monograph (fig. 40). In any event, one of his students apparently suggested that he should commission and publish a whole series of such scenes. According to Unger himself (text 31), he was at first hesitant to do so. He gave as his reason what we have suspected in the similar reluctance of Cuvier, De la Beche, and Buckland: he was afraid of "wandering from the actual domain of science, into the realms of imagination," in other words, of being thought unduly speculative. However, after consulting his fellow botanist Karl von Martius (1794–1868) in Munich, Unger approached a well-known landscape painter in Graz, Josef Kuwasseg (1799–1859). He was so pleased with the artist's trial efforts that he overcame his worries about his scientific reputation. Indeed, he and Kuwasseg now formed a partnership as intellectual and artistic, if not financial, equals. The team of scientist and artist embarked on the most ambitious project of its kind yet undertaken; and Unger's introduction to their work is unstinting in its appreciation of Kuwasseg's artistic achievement.

The Primitive World in Its Different Periods of Formation (*Die Urwelt in ihren verschiedenen Bildungsperioden*, 1851) was published as a magnificent folio atlas, containing fourteen large lithographed scenes by Kuwasseg, accompanied by an explanatory booklet. Unger's German text was translated into French by his Alsatian (and therefore bilingual) colleague Wilhelm Philipp Schimper (1808–80), the professor of natural history at Strasbourg; with the two texts printed in parallel,

Text 31

ORIGIN AND DESIGN OF THE WORK

At the conclusion of a course of geological instruction which I was called upon to give to a few Students, a wish arose in the mind of a person, no less amiable than gifted, that our labours might be rendered permanently useful by the circulation of drawings illustrative of the subject which had engaged our attention.

It is impossible to deny that this method possesses the merit of setting forth geological science in such a manner, as to render it easy of apprehension to those who have neither leisure nor inclination to occupy themselves with its details. This persuasion, and the desire of being useful, overcame the diffidence I originally experienced at the prospect of wandering from the actual domain of science, into the realms of imagination.

I felt, however, that difficulties almost insurmountable obstructed the execution of this project; the attempt was consequently delayed, until the inspection of some experimental drawings, submitted to me by the talented artist Josef Kuwasseg, not only convinced me of the possibility of levelling gradually all obstacles to the success of the undertaking, but induced a hope that these representations of the primitive world might be found not wholly to lack that mysterious charm which belongs to the contemplation of the distant past, and to the memory of our dreams.

Such was the origin of a series of drawings which exhibit, in order, the periods of the Earth's formation: but desires not unusually increase in proportion to the frequency of their gratification, and a wish was expressed that the opportunity of procuring landscapes originally designed for the use of a few individuals alone, should be extended to the public.

I would not omit to tender here my thanks to Messieurs Bernard Cotta, Hermann Meyer, Et. Endlicher, von Tschudi, Fitzinger, and the other men of learning to whom I am indebted for advice and corrections, and by whose researches I have not scrupled to profit.

But it is to the Artist himself that my gratitude is chiefly due. Unwearied by frequent trials, he finally attained such a perfect comprehension of the conceptions I had formed of these remote periods, that the undefined visions of my fancy were, by his genius, developed into clear and vigorous images.

I can claim, then, as my part in the performance of this undertaking, only the communication of my thoughts and conjectures: if these have given inspiration to the Artist, if to these is attributable the existence of his productions, not to me, but to him with whom the scheme originated, belong alike the merit and the praise.

Franz Unger, *The Primitive World in Its Different Periods of Formation* (1851).

NOTES. The colleagues whom Unger acknowledges were Carl *Bernhard Cotta* (1808–79), professor of geology ("Geognosie") and paleontology at the famous mining school at Freiberg in Saxony; Christian Friedrich *Herrmann von Meyer* (1801–69), a Frankfurt bureaucrat and, in his spare time, a prominent paleontologist; *Etienne Endlicher*; Johann Jacob *von Tschudi* (1818–89), a naturalist who had traveled extensively in South America; and Leopold Joseph *Fitzinger* (1802–84), a Viennese physician and zoologist.

the work was linguistically accessible throughout the scientific world.[2]

The title was carefully worded: it takes the popular term "primitive world" (*Urwelt*), which suggests an undifferentiated deep time, but juxtaposes it with a phrase that emphasizes the *different* periods of earth history. Unlike the scheme that Buckland had suggested privately many years earlier (text 15), those periods are now illustrated in true geohistorical order, from oldest to youngest, allowing the reader to trace the progress of life toward the present. The resultant similarity to the traditional sequences based on the biblical narrative of Creation is probably no coincidence. Unger's cultural environment was deeply Catholic: three of his subscribers were high ecclesiastics, and he himself bore the name of a great early Jesuit, Francis Xavier. He is likely to have been familiar with sequences of biblical illustrations: if not with Scheuchzer's (figs. 1–7), then at least with those in the widely popular *Pictorial Bible* (*Bilder-Bibel*, 1836), which was designed explicitly for German-speaking Catholics.[3]

Not surprisingly, Unger's selection of periods reflects the succession of major formations of strata that were most familiar to geologists and their students in central Europe. Since he was a botanist, it is also not surprising that, in contrast to most of the earlier scenes that have been reproduced here, Unger gives much greater prominence to the plant life than to the animals. Finally, his landscape painter makes no attempt to follow De la Beche's innovative lead (fig. 19) in taking the reader underwater for a fish's-eye view of the ancient marine world. Conversely, however, Kuwasseg's finely composed scenes, which reflect contemporary traditions of landscape painting, do import a believable *human* viewpoint into deep time, and thereby make it imaginatively accessible. Only in one stormy moonlit scene (fig. 48), and perhaps in the only scene that focuses on animals rather than plants (fig. 51), is there any hint of the nightmarish world of John Martin.

In his brief introduction (text 32), Unger adopts a decidedly defensive tone, evidently to forestall the anticipated criticisms of his academic colleagues. He stresses the difficulties of reconstructing scenes from fragmentary fossil remains and the provisional nature of what he has done. However, he also goes on the offensive, criticizing previously published scenes (it is likely that he had Goldfuss's in mind). Such scenes, he argues, have been highly artificial assemblages of all the fossils known in certain strata, whereas Unger's scenes—or rather, Kuwasseg's—portray prehistoric landscapes, and their animals and plants, with much greater naturalism.

One further general point needs to be made about Unger's explanations of Kuwasseg's scenes. Unger was primarily a botanist, not a geologist. For his general understanding of the physical history of the earth, he adopted without question the consensual view of most geologists around mid-century: namely, that the earth has experienced a *directional* development from its earliest beginnings up to the present, as much in physical and chemical conditions as in the character of plant and animal life at its surface.[4]

This explains Unger's frequent allusions to the alien environments of the plants in his earlier scenes, and the gradual approach toward a more familiar—and humanly congenial—state in the later scenes. For example, Unger assumes that the temperature has decreased steadily and that climates have become more temperate throughout earth history, because the earth has been cooling down from an initially hot state. Likewise, he takes it for granted that the oceans were at first almost

Text 32

INTRODUCTION

I have undertaken to give in these landscapes a pictorial representation of the great geological periods which have passed in succession, from the time when the surface of the earth was animated by the first organic beings, to the era of man's creation. They have been executed according to the facts already established by geology and palaeontology, such as the plastic nature of the earth, as taught by the former science, and the constitution of vegetable and of animal life according to the latter. With this design, I have employed all the data which indicate with accuracy the character of the strata distinguishing each period, the manner of their destruction and of their re-formation in new strata, the influence of the earth's interior upon its surface, and the general distribution of land and water. I have also examined attentively the numerous remains of plants and animals which have been imbedded in the sedimentary deposits of the primitive seas; for, revealing to us what were the flora and the fauna of each period, they teach us also with the greatest accuracy the nature of the climate proper to them. Unhappily the incomplete state in which these plants and animals are mostly found, has compelled us, in reproducing them, to trust sometimes to the aid of analogy for supplying those parts, which, were we to limit ourselves to an exact copy, could not be depicted. Notwithstanding, however, the scrupulous use that has been made of this analogy, it is probable that some of these representations may not quite answer to their true types, especially those of the remote ages, which exhibit no point of resemblance to the world in its present state. These landscapes, then, are compositions in which have been collected and grouped images of the various phaenomena of primitive nature; we confess, however, that they may wander from the truth where these phaenomena have not left a perfect impression. As, in the present day, the landscapes of our painters are but seldom exact and servile copies of natural scenery, so these drawings of the primitive world give only the general character, and not the precise reproduction of ancient periods; and they may be regarded as combinations made by an artist, who, having travelled in these remote times, has on his way collected the details, to group them, at his pleasure, in a single view.

It is with regard to its aesthetic bearing alone that this work must be considered. What has already been done in this way does not attain the end which I have designed to reach; for the collections of drawings in which has been restored to the remains of fossil plants and animals that which time had rendered incomplete,—collections made according to the system of formations, or to the date of the appearance of these beings on the earth,—cannot by any means be looked upon as natural landscapes. The drawings of some partial periods of palaeontology that are given in certain geological works are even less worthy of the name, for in them are accumulated the greatest procurable number of types, brought together rather with regard to an exact chronology than to a perfect representation of the various primitive countries.

I have sought, then, to direct attention to some scenes of those great epochs, rather than to instruct by the mere representation of a number of individual objects. My design was to show the effect of these landscapes of the ancient world, and it was chiefly to the talent of the artist that the attainment of our end was entrusted.

Severe critics will probably object to many parts of this scheme: botanists may say that the majority of the plants are idealized; and that it is even probable, judging from the imperfect knowledge we have of them, that their form and general appearance may have been totally different from what is here depicted: zoologists may object that the great vertebrated animals which occupy a prominent place in many of these pictures, do not always correspond with their types; while geologists will perhaps opine

that the character of the earth's surface has been but too often conceived from a collection of general ideas;—in a word, it may be thought that the appearance which I have given to this ancient world cannot boast the truthfulness that was expected from it, and that my labour bears the impress rather of an arbitrary and capricious imagination than of a searching and correct observation of facts. If I gave each of these landscapes as a perfect reproduction of an actual appearance, these objections would have weight; but my desire has been to express only a possibility which should be nearest to the truth, in the cases where sure data are given by geology and palaeontology.

It may be further questioned whether a work into which enters perhaps as much of error as of truth, may not tend rather to retard than to advance the progress of science. But these landscapes will not lose all their worth in the eyes of those who know how much of progress is owing to certain hypotheses, and what great support they have in all ages afforded to the feeble infancy of the natural sciences.

If these representations, in uniting a thousand observations in a single point of view, induce a more facile appreciation of the whole—if they offer to any who are little learned in the science an interest such as to create in them a desire to accomplish the study—if they awaken in men of liberal education a taste for the contemplation of these long-banished periods, and cause them to look upon the Present as the result of this great Past—they may at a later period give place to more enduring works, and be thrown aside, like dried leaves fallen from a tree, whose former verdure is a matter of astonishment.

Franz Unger, *Primitive World* (1851).

global in extent, with no more than a few scattered islands, and that only much later did extensive continents develop. Even then, he assumes, their relief was at first slight, so that high mountain ranges are a relatively recent feature. Conversely, he maintains that volcanic activity has in general declined in the course of time, and with it the concentration of noxious gases in the atmosphere. More specifically, he attributes—as did most geologists—the profusion of plant life in relatively early geological periods, such as the Coal or Carboniferous period to a higher concentration of carbon dioxide: the atmosphere, like every other feature, had changed directionally toward its present state.

The first scene (fig. 43; text 33) represents the earliest geological period for which Unger felt there was sufficient fossil evidence to warrant any reconstruction at all. The "Transition" formations, underlying and therefore older than all the Secondary ones, were still very poorly understood, since they were generally much disturbed or altered and rarely contained abundant or well-preserved fossils. In fact, Unger's first scene shows simply an impoverished version of the Coal forest scene that follows. At this early period, there were only a few far-flung islands in a vast tropical ocean. The landscape is scattered with tree-sized cryptogams taking the place of modern palm trees. The only component that needs special comment is the curious *Stigmaria* plant in the foreground, which had also figured in Goldfuss's scene (fig. 40). Later discoveries revealed that this common fossil was the root system of a tree-sized plant; the two are rarely preserved attached. Unger, in common with other fossil botanists of the time (as well as Goldfuss), made it into a bizarre "dwarf plant."

The second scene (fig. 44; text 34) is based on the Coal Measures of the Carboniferous, the strata in which most of the industrially important coal seams were found. It

Text 33

In the boundless ocean which overspread the terrestrial globe, appeared at first a few small islands separated by wide intervals, the earliest land.

The first landscape [fig. 43] presents to our view one of these primitive groups, of a rocky, probably of a granitic soil; not yet elevated into mountains, but forming plains which surmount steep and irregular coasts. The greater number of these plains have already existed for a long period, but in the distance we see one which has arisen only lately; and thick vapours ascend from its mass, which has as yet been but imperfectly hardened by contact with the waters. The atmosphere, filled with these vapours, covers with dark clouds and thick fogs these solitary isles. Owing to the thinness of the earth's solid crust, the high temperature of the globe permitted scarcely any activity to exist but that of inorganic nature, such as the action of affinities in the formation of solids and fluids, and great physical phaenomena. But as soon as the beams of the sun had forced a passage to the earth's surface, a new kind of existence began to animate the waters and these hitherto desert isles, in the various forms of animal and vegetable life. These first-born of the earth had a simple organization, but they exhibit the germ of the riches of future creation.

We leave the waters and their inhabitants to notice the appearance of the solid earth. Here we remark with astonishment vegetable life presented in forms which are utterly foreign to us; gigantic trees bearing the simplest foliage, or leafless plants with great cylindrical stems, or trunkless trees with verticillated branches. These vegetable existences, sometimes woody, but mostly herbaceous and fleshy, bore neither fruits nor flowers, but in their stead sporules; they were, in a word, *Vascular Cryptogamia*. The only plants of this period whose appearance bears a striking resemblance to any now produced are Ferns, of which there are several groups in this picture: some covering the naked land, others hanging by their roots to the bark of trees, while their

Figure 43. "The Transition Period":
drawing by Josef Kuwasseg,
lithographed by Leopold Rottman,
from Franz Unger's *The Primitive
World in Its Different Periods of Formation* (1851).

delicately-formed leaves droop towards the ground. But the most extraordinary plant of this landscape—extraordinary alike in the majesty of its greatness and in the singularity of its appearance—is the *Lomatophloyos crassicaule*, a tree more peculiarly proper to the following period, though not unknown to this. Its spirally-disposed branches, which have for the most part fallen from the lower portion of the trunk, and now spring only from the top, bear at their extremities alone a thick mass of linear and carnose leaves, placed on a scaly tuft. Upon the fall of the leaves these tufts are discovered, and give a singular appearance to the tree.

Another great tree is the *Sigillaria*, a genus rich in species. Its simply organized and undivided trunk bears a majestic and tufted crown of long and drooping leaves. The furrowed bark is marked with regular series of impressions left by the former foliage. Various specimens of this genus, in their different stages, are grouped in the first landscape.

In the distance are trees with verticillated branches, and a foliage of great delicacy: they belong to the *Calamites*, a genus including great variety of form. But that which strikes us most forcibly in this landscape, and we may almost say that which forms its distinguishing feature, is the presence of those marshy plants which cover the water of the foreground so thickly, the *Stigmaria ficoides*. Their short stems rising but slightly above the surface, are divided into long dichotomous branches which bend downwards, submerging ordinarily their extremities, which bear long, thick and rounded leaves. These dwarf plants occupy a vast area, and present the appearance of a cupola crowned with young and upturned branches. They flourished chiefly in the more shallow lakes; but they extended, doubtless, to where these lakes mingled with the salt waters of the ocean. All these plants existed under forms more or less similar until the next geological period. A very few are found which are peculiar to the Transition period, but they are of a most remarkable organization. These plants are straight, cylindrical, almost

always bare of branches, of fruits and of flowers, but extraordinarily pulpous, betraying, notwithstanding their internal woody fibre, a simple structure: they may be justly regarded as the earliest terrestrial vegetation. It is scarcely possible that creative power should have been exercised on the structure of less complicated forms than those which border the marshes, such as the *Didymophyllon Schottini* and the *Dechenia euphorbioïdes*. A little freshwater herb, the pretty *Annularia fertilis*, which appears between the tufts of the *Stigmaria*, serves to diminish the melancholy and singular appearance of these plants.

Franz Unger, *Primitive World* (1851).

Text 34

We are here [fig. 44] transported to the interior of a damp forest intersected by stagnant waters, and situated in a small island. All proclaims one of those primitive lands which have not yet been inhabited by living beings, and which are subject to the laws of inanimate nature alone. The light of the sun obscured by the foggy atmosphere, penetrates but feebly through the interwoven branches of the trees; and the dense and humid air increases the apparent distance of all the surrounding objects. The trees which predominate here are the Lepidodendrons, characterized by their rough bark and by their dichotomous branches; there is a group of these in the foreground. Their branches, of vigorous growth and thick foliage, recall, as well by their form as by their fruit, our own fir-trees; but they give a shade yet deeper and more mysterious. A considerable number of parasitical plants have grown alike on the healthy and on the decaying trunks, and are instrumental in heightening the picturesque effect of the landscape by adding to its strange and primitive appearance some points of resemblance to our own scenery.

In the distance appear among the Lepidodendrons some Ferns, easily recognized by their thin tall

Figure 44. "The Coal Period: First Scene," from Franz Unger's *Primitive World* (1851).

is thus directly comparable to Goldfuss's second scene (fig. 40), but the design is quite new. By not showing the foliage in close enough detail for precise identification, Unger avoids Goldfuss's problem of relating specific forms of foliage to specific types of trunk. Consequently his scene is a true landscape, and less an assemblage of museum specimens. Kuwasseg conveys very successfully the atmosphere of a gloomy tropical swamp. Unger remarks that the land was "not yet inhabited by living beings." In context it is clear that this refers only to terrestrial animals: he was well aware of the abundant fossils of marine animals in strata older than the Coal, although he chose not to illustrate any of them. Of course what he did illustrate was a world full of plants.

The third scene (fig. 45; text 35) presents, uniquely, a second view from the same geological period. It illustrates a torrential tropical storm scouring a channel through the peaty accumulation of a future coal seam, creating one of the interruptions that were all too familiar to miners.

The fourth scene (fig. 46; text 36) is of a period later than that of the Coal, represented in central Europe by a major sandstone formation with associated volcanic rocks (*Totliegende*).[5] Prominent in the background, and in Unger's text, are recently erupted domes of volcanic rock ("porphyry"); this volcanic activity is held responsible for climatic conditions that restrict the flora to far less varied forms than before. The implication, reflecting a view commonly held by geologists at the time, is that this period was characterized *globally* by such disturbances and the climate they caused.

The fifth scene (fig. 47; text 37) illustrates the period of the "Bunter Sandstone," the lowest and therefore oldest of three major formations widespread in central Europe, known collectively (for just that reason) as the "Trias." The plants, though less profuse than those of

continued on page 112

stems, no less than by their graceful groups of foliage drooping from the summit. They belong to the genus *Cyatheites* (*Pecopteris*). A young stem belonging to an individual of the same genus is seen to the left in the foreground. The parasitical plants which we have before remarked, cling also to the family of Ferns. Those which form garlands suspended from the branches, are of the genus *Hymenophyllites* and *Sphenopteris;* those which cover entirely the trunks with their larger leaves, are *Woodwardites* and *Trichomanites;* and the others with thick pendents clustered here and there, are *Cyclopteris* (*Adiantites*).

In the background, nearly hidden, is a group of *Calamites*, those gigantic plants of the ancient world which contributed not a little to the singular appearance of that vegetation. Finally, we must mention a little aquatic plant with ring-shaped leaves and branches belonging to the genus *Sphenophyllum*, so rich in species; this singular form of *Asterophyllites* has not been preserved to our own times.

Franz Unger, *Primitive World* (1851).

Text 35

While the preceding Plate [fig. 44] represents the gloom of a primitive forest of evergreen trees, in which the changes of spring and autumn are never seen, in that which is before us [fig. 45] it appears that Time has arrested his steps to open for our perusal the book of the history of anterior periods, which may be deciphered in the accumulated vegetable remains whose uninterrupted succession connects the present with the past. It is no longer doubtful that the massive strata of combustible minerals which we call Coal, owe their origin to an accumulation of ancient vegetables which have grown one upon another; of this growth we have an inferior example in our own peaty strata. With us, decayed leaves and branches serve, immediately after their fall, to form the soil for the next generation of plants; whereas, in these primitive ages, Death set his seal for myriads of years on the plants which accumulated on the earth, and which, in lengthened

Figure 45. "The Coal Period: Second Scene," from Franz Unger's *Primitive World* (1851).

course of time, have lost completely their pristine physiognomy; not even the faintest trace of their original organization being discoverable. These great vegetable masses, which increased in the course of ages to a thickness of several fathoms, and extended over very large areas, have been disjoined and partially destroyed by the shocks which have disturbed the earth's crust. In the Plate, we see a portion of the primitive forest, into which we have penetrated in the preceding drawing, resisting the ravages of the waters by its colossal plants heaped together, and by its amassments of peaty substance; but the general inundation appears ready to overwhelm it in its turn, and to cover it with a deposit of mud and sand. Here, as in the preceding Plate [fig. 44], we remark trees, in which is traceable but a remote analogy to any that exist in these times, and whose types belong exclusively to the ancient world. The principal groups are formed by the Lepidodendrons and by the Sigillaria, equally numerous in this period. Two trees of the latter family rise to the left of the foreground, and bend before the violence of the tempest. Opposite, we see, at the edge of the overwhelmed land, some arborescent Ferns, whose light leaves, agitated by the wind, serve, like wings, to sustain them above the waves. In the distance, transparent crowns of Calamites escape, through their flexibility, destruction by the hurricane which roots up and forces to the earth the Lepidodendrons, whose solid trunks and heavy foliage resist instead of bowing to its fury.

This scene of ruin is made visible only by the electric flash which sweeps across the thick clouds. This rich and strange vegetation, deluged by torrents of rain, existed under so burning a sky, and on a soil so softened with water, that the fierce heat of a tropical climate and the damp atmosphere of the Chonos isles [off the coast of Chile] are inadequate to convey to us an idea of the intense extremes which prevailed.

Franz Unger, *Primitive World* (1851).

Text 36

The period of the Red Sandstone, distinguished by very numerous disturbances of the earth, could not be favourable to the development of vegetation. Great solid masses alone, such as vegetable trunks or their remains, were able to resist the force which constantly changed the condition of the earth. We find, consequently, buried in the strata formed in this period, only fossils of that nature, or but seldom leaves and fruits; and those even which escaped total destruction owe their safety to the shelter of their tombs. This period, however, was not utterly destitute of plants, of which some kinds were even very elegant, though they cannot sustain a comparison with the richer vegetation of the anterior formations. This period, then, has also its vegetable physiognomy, in harmony with the constitution of its soil and atmosphere. We find ourselves in the midst of one of the sublimest scenes of nature [fig. 46], on the gigantic wreck of a mountain recently destroyed by an eruption. The ejected prophyry has forced a passage across the ravines, and formed itself into conical cupolas on the wrecks it has produced. This soft and muddy mass could not but be disposed in forms more or less rounded; thus we see in the midst of the landscape one of these porphyry domes in its earliest stage, before time has furrowed it with clefts and precipices. It is naked, without the least trace of vegetation, for the rock is yet imperfectly hardened. To the left, at the bottom of the valley produced by the eruption, appears a mountain of the same formation.

The foot of these recently-formed peaks is surrounded by a mass of stony fragments thrown up in the late disturbance. From clefts still open, rise vaporous clouds,—evident proofs of the volcanic operations which agitate the depths of the earth; they foretell the probable renewal of eruptions, or are perhaps but the latest manifestations of a strength well-nigh exhausted. A column of smoke, attended with miry eruptions, rises from the surface of the sea, and bringing with it a considerable body of liquid, forms a fountain of prodigious height. This

Figure 46. "The Red Sandstone [Permian] Period," from Franz Unger's *Primitive World* (1851).

steaming water cools quickly in the high regions of air which it reaches, and falls in showers near the spot whence it rose. Terrible lightnings pierce incessantly these heated vapours, the sea around is greatly agitated, and this volcanic phaenomenon has a character no less sublime than terrific. The excessive dampness of the air, an inevitable consequence of these atmospheric phaenomena, nourishes a vegetation, which, though stinted, is distinctive, forming a striking contrast with that of the foregoing period. The landscape is adorned with the upright stems and elegant foliage of the *Psaronius;* young plants of this genus, and ferns having rhizomatous tubercles, grow between the fragmentary rocks. To the elegance of the *Psaronius* is added that of the *Cycadeae,* a family unknown until this period. There are several groups on the coast between the blocks of stone, and in the distance. Their tufts of leaves are transparent and full of grace. This kind of foliage has entirely disappeared, but the trunk of the *Cycadeae* is still found in the genus *Medullosa.*

Franz Unger, *Primitive World* (1851).

Text 37

A period infinitely less turbulent, that of the Upper New Red Sandstone formation, followed the destructive shocks which had signalized the preceding geological period; but, owing to the remains of the metalliferous vapours of the Zechstein [formation], few terrestrial or even marine plants could be preserved. The deposits of sand, clay and marl spread over the bed of the sea, then surrounded by lowlands, were so slightly disturbed that their alternation is regular, and their general appearance proclaims a tranquil age, favourable to the development of organic life, and especially capable of producing a rich and varied vegetation. There is, in fact, in this period great progress visible in vegetable life, which though not so thick as that of the coal formation, attains more perfect and more elegant forms. The remains of this vegetation, which the waters carried

away and buried in their deposits of mud, have been preserved in a state of such completeness, as to have afforded us all the materials of the landscape now under contemplation [fig. 47]. It transports us into the midst of a low country surrounded by shallow waters; in the distance are islands formed of water-flowers. On masses of Red Sandstone, but little elevated above a mud-bank, rises a forest of thinly-scattered trees; they are the Coniferous plants which distinguished so eminently this period. To the left of the foreground are some old trunks of *Haidingera speciosa;* further to the right and at the back is a group of *Voltzia.* Branches covered with leaves and bearing cones have been preserved whole, and show us that the appearance of these trees bore a great resemblance to that of the *Araucaria,* especially the *Araucaria excelsa* of Norfolk Island [north of New Zealand]. Below the Conifers appears here and there a tree (the *Yuccites Vogesiacus*) ramified only at the summit, and having tufts of leaves similar to those of some Monocotyledons, such as the *Alethris* and the *Yucca.* We exhibit also, as characteristic of the Sandstone period, the two herbaceous plants which partially cover the forest or extend over the marshy foreground. In the former, the remarkable *Aethophyllum speciosum,* are united the different types of the *Lycopodiaceae* and of the *Typhaceae;* and the latter, the *Schizoneura paradoxa,* belongs to the *Equisetaceae* rather than to the *Smilaceae,* to which it has been hitherto referred. In the centre of the foreground rises a beautiful *Cycadea,* the *Nilssonia Hogardi,* among the Ferns, growing on rotted branches of trees or in the clefts of stones, and representing the earliest state of the fossils called *Crematopteris typica, Alethopteris Sultziana, Neuropteris elegans,* etc. The animal and vegetable world of this period display a most interesting connexion. We have, however, only faint traces of beings superior to the Fishes which lived at that time, though we cannot doubt that the earth was peopled by amphibious animals, and even by problematical beings partaking in some degree of the nature of Birds. Our most cer-

Figure 47. "The Period of the
Bunter Sandstone [Lower Trias],"
from Franz Unger's *Primitive World*
(1851).

the Coal period, are said to be of "more perfect and more elegant forms." Conifers are now abundant, and other forms such as cycads are also noted. Even at this rather late stage in the whole history of life, Unger is cautious about attributing any animal life to the dry land; but in this scene the fossil evidence known to him does at last justify showing a large amphibian crawling toward the shore.

The sixth scene (fig. 48; text 38) represents the "Muschelkalk," the limestone of marine origin that forms the middle third of the Trias (even anglophone geologists still use its German name). Unger interprets this period as a temporary reversal of the general trend toward an increasing area of land: the sea has advanced, to his evident regret, since he is reduced to showing his beloved plants as drifted wrecks. To portray the marine life, he uses the time-worn device of a low tide, which exposes some coral reef, stranded mollusc shells and plantlike crinoids. For the first time, the most prominent organism is an animal, namely, the large reptile *Nothosaurus*. It is not clear why Unger (or Kuwasseg) chose moonlight for this scene, unless it was to create a "desolate" effect to match Unger's response to an almost plantless phase. In any case, the result is nearer in feeling to Martin's style than any other of Unger's scenes.

The seventh scene (fig. 49; text 39) brings a welcome return to a terrestrial world; it portrays the time of the "Keuper Sandstone," the uppermost and therefore youngest part of the Trias; like the Bunter, this formation was considered to be of nonmarine origin. A landscape of low hills and marshes runs down to a large lake in the background. The most prominent plants are those related to modern horsetails, some of them the size of trees. Crawling purposefully across a patch of sandy ground is the amphibian *Labyrinthodon* (so named on account of the complex structure of its teeth). The English

tain proofs are the footmarks made on the wet sand of these flats. Some of these imply gigantic proportions; they are those of the *Cheirosaurus*, an amphibious animal, and of a winged biped, having produced the *Ornithichnite*. The animals of the genus Salamander which are here represented, may give an approximately just idea of the enormous amphibious animals which inhabited the primitive forests, and augmented the number of their remarkable features.

Franz Unger, *Primitive World* (1851).

Text 38
The flat country that had been gradually freed from the dominion of the sea, and covered with a diversified existence, has again been invaded by the ocean. A boundless sea has overwhelmed the flats adorned with vigorous Conifers, the marshes covered with singular Equisetaceae, the reed-forest glades in which lived extraordinary monsters, amphibious or biped. In their place, the waters have afforded an asylum to new colonies. We see no longer those great vegetable productions increased to gigantic proportions in the humid and calm air which surrounded them, but aquatic animals with forms resembling those of plants, and bearing flower-like heads, with myriads of Mollusca and Crustacea. Small islands in the distance alone recall the preceding periods by some scanty remains of vegetation which appear upon them. Here [fig. 48] we discover the desolate banks of one of these tracts of land, surrounded on all sides by the rolling waves;—a serious and melancholy scene, on a broken soil, at once the erection and the habitation of Corals and other Zoophytes: exposed to view by the ebbing tide, we surprise these beings of the marine world in the secrets of their mysterious existence. By the side of the marvellously formed shells of the *Ammonites* (*Ceratites nodosus*) are seen other Mollusks, the *Nautilus bidorsatus*, and beyond, the *Pecten discites*, the *Plagiostoma striatum*, *Turritellae* and other Testacea, all belonging to the

112

Figure 48. "The Period of the
Muschelkalk [Middle Trias]," from
Franz Unger's *Primitive World*
(1851).

Muschelkalk formation. The beautiful *Encrinites liliformis,* with their jointed stems, have been borne here by the waves from the depths of the sea. A singular monster of the Crocodile kind, having borrowed its large fins from the Cetacea, contemplates with eager eye the marine animals which it destines for its prey. It is the *Nothosaurus giganteus* attempting to mount these coral rocks in order to secure its food. To the right of the background we see, occupied in the same pursuit, two monsters of smaller proportions (*Nothosaurus mirabilis*). Remnants of the trunks of trees which rise here and there like skeletons above the water, announce the proximity of land. A tempest has driven them thus far, as may be clearly perceived by the remains of other plants which still cling to them. These trunks belong, doubtless, to a great Conifer, the *Pinites Goeppertanus,* probably differing but trivially from the *Voltzia;* and the smaller plants may represent to us the problematical *Endolepis,* mixed with the fern *Neuropteris Gaillardoti,* and with the sea-weed which has been tossed ashore (the *Spherococcites Blondowskianus*). The melancholy aspect of this view, an approximately faithful representation of the appearances in this period, is augmented by the darkness of the night. The moon which illumines the desolate scene gives not her usual soft and silvery light, but throws over it the pallid glimmer of a distant star.

Franz Unger, *Primitive World* (1851).

Text 39

The character of the earth's surface is little changed since the New Red Sandstone period, although considerable strata have been formed of calcareous deposits containing prodigious remains of shells that the Muschelkalk had left behind. The deposits made during these two periods formed a soil of small extent, without mountains or even hills; here the vegetation of the Keuper had its origin. The flat and level banks of a large lake are extended before us [fig. 49]. The dry and sandy downs are still completely bar-

ren; vegetation has been able to establish itself only on the marshy lowlands that the water fertilizes. On this damp soil we observe a vegetation rare indeed, but not remarkable for richness. To the right we perceive the borders of a forest composed of *Calamites arenaceus,* trees prevailing in this period. Their furrowed trunk rises without branches to a considerable height, and bears a cupola of leaves gracefully disposed on light drooping branches. Some herbaceous plants of our own time are somewhat similar to these trees, which bore no fruit like the great inhabitants of our forests: these existing plants are the Equisetaceae (Horsetails), of which the *Equisetum sylvaticum* is the most majestic. On several of the trunks in the foreground we note a climbing plant, whose oval leaves are borne on long stems, and whose fruit forms bunches of berries; it is the pretty *Preisleria antiqua,* of the family of the *Smilaceae,* of which it is the earliest type. Near this plant are several fine kinds of Ferns which still enrich the forests: among others the majestic *Anomopteris Mougeotii,* characteristic of the Upper New Red Sandstone. The marshy soil which forms broad belts to the left of the drawing is entirely different in character. The most conspicuous plant in this region is the *Equisetites columnaris,* whose long stem rises above all the others, bearing on its summit fruit of an oval form. This tree, the most gigantic of its kind either in the primitive or in the modern world, appears to have been herbaceous and of tender growth, so that its existence was necessarily of but short duration. The marshy plant most frequently encountered in the Keuper is a kind of rush, which is represented in the foreground, and which grew to the height of about six feet; it is the *Palaeoxyris Münsteri.* At the furthest extremity, on a slight elevation, some *Cycadeae* are growing; among which is the *Pterophyllum Münsteri.*

The contemporary animal world consisted chiefly of testaceous Mollusks, Polypi, Crustacea, and other fish peopling the waters: on dry land it is probable that there were Amphibia alone. The animal which we here remark, and with whose form we are ac-

Figure 49. "The Period of the Keuper Sandstone [Upper Trias]," from Franz Unger's *Primitive World* (1851).

quainted only by a few bones and teeth which have been discovered, and by footmarks made in the plastic soil, is the remarkable *Labyrinthodon pachygnathus*. This animal was probably nourished by shell and other fish, and appears to have inhabited the marshy plains of this period. The damp, warm atmosphere of this age was in perfect harmony with its animal and vegetable productions. The clouds which covered the earth were so thick, that the sun's rays could reach it only by the intervals which existed for a moment between the masses of moving vapour.

Franz Unger, *Primitive World* (1851).

zoologist Richard Owen (1804–92) had boldly reconstructed this creature from a very few bones and teeth,
and had even more boldly inferred that its large quadrupedal footprints were those found commonly in the
Keuper in many parts of Europe.[6] In Kuwasseg's scene,
its footprints are shown, rather curiously, in front of the
animal; presumably it had been that way before!

The eighth scene (fig. 50; text 40) represents the period of the Oolite (in Unger's terms, this includes the
English Lias and corresponds to the French *terrain jurassique,* which in turn yielded the modern English name
Jurassic). The contrast with earlier scenes of Liassic life is
striking, but again this simply reflects Unger's botanical
emphasis. It is the trees that are once again most prominent. Unger's reconstructions from fragmentary fossils
are in some cases just as bold as Owen's labyrinthodon or
Mantell's iguanodon: here, for example, he admits that
as evidence for the pandanus trees with their stiltlike roots
he has nothing but fossil fruits. The ichthyosaurs that
have so dominated earlier representations of the life of
this period are reduced ignominiously to a single beached
skeleton; the plesiosaurs to one distant figure; and the
pterodactyles to a few equally distant birdlike forms.

The ninth scene (fig. 51; text 41) brings animals into the
greatest prominence that they ever enjoy in Unger's series.
It represents the period of the Wealden formation of
southern England, which Mantell's research had made familiar internationally. Abundant ferns and cycads domi-
nate the vegetation, while the Clathraria tree signals the
arrival of the monocotyledons. The luxuriance of the
plant life is attributed explicitly to the still high proportion of carbon dioxide in the atmosphere.

Pride of place, however, is given to Mantell's iguanodons. Their portrayal suggests that Unger and Kuwasseg
had available to them both Richardson's scene and that
drawn by Martin for Mantell (figs. 34, 35). The individ-

Text 40

In the eighth design [fig. 50] is presented to our view
the vast ocean which continued during the Oolite
period to cover the greater part of the earth's surface, and above which rose small islands, where were
visible the upheaved strata of the Trias formation.
One of these maritime lands is here pictured, with its
vegetable growth, its singular inhabitants, and its en-
circling mass of Coral, which, though yet but little
elevated above the sea-surface, is of great breadth all
along the coast. The cloud-covered sky, and the lightnings which flash across it, indicate a burning atmosphere, which is declared with equal clearness in the
physiognomy of the vegetable and animal worlds. If
we examine the vegetation, which here at least is sufficiently luxuriant, we shall see that plants belonging
to the family of the *Cycadeae,* especially the *Pterophyllum* and the *Zamites,* form its distinctive feature.
They are no longer seen growing singly and scattered, but they now cover large areas with their thick
groups. The great trunk which appears in the middle
of the drawing belongs to the *Pterophyllum.* Covered
with immense leaves, which grow on little thick
gnarled branches, it rises perpendicularly to the
height of many feet. Trees, in which the characteristics of the *Zamites* cannot but be recognized, occupy
the spot which is furthest distant from the forest,
and are partially hidden by the large foliage of the
Pterophyllum. Their trunks, encircled by the marks
which have been left by the leaves of former years,
and which recall to the mind the stumps of the *Cy-
cadoidea megalophylla,* are crowned with a thick
cluster of leaves more than six feet in length, and
bear a cone-shaped fruit of great size. Bushes belonging to various species of the same genus cover
the ground and form the underwood of the great
forest. In the foreground of the picture, and near
the centre, is a plant, which, by its short gnarled
trunk, and by the singular form of its foliage, is
known to be the *Zamites undulatus.* To the extreme
right, some trees of the Pandaneae family give a distinctive character to the vegetable physiognomy of

Figure 50. "The Oolitic [Jurassic] Period," from Franz Unger's *Primitive World* (1851).

this period by the singular position of their roots, and by the beauty of the foliage which adorns the extremity of their branches. Neither its leaves nor its roots have been hitherto discovered in a fossil state; only the ball-shaped fruit (*Podocaria Bucklandi*), from which, however, the whole conformation of the plant has been ascertained with perfect accuracy. Elegant Ferns, the *Cyatheites obtusifolius*, the *Cyatheites acutifolius*, *Pecopteris* and other kinds, cover at intervals the land, or appear between the crevices of the sundered rocks. Of these, the most remarkable (the *Hemitelites Schouwii*) display their beautiful blossoms under the *Pterophyllum* of the foreground; and the smaller species of *Sphenopteris* and *Hymeno-phyllites*, which are scarcely to be found, must not be overlooked. The distant land which extends to the horizon on the left, is covered with a similar growth consisting of Cycadeae, Ferns, Pandaneae; with these are mingled some Coniferae of the genus *Thuites*, as we are convinced by reference to the plants which grow on the margin of the bank on the right of the drawing, and which may be more easily recognized from their closer proximity to the spectator. The trees which we perceive on the deposits left by Polypi, are not here of the *Pterophyllum* kind, but several species of Palms, whose fruits, discovered in a fossil state, prove beyond the possibility of doubt, the existence of those trees in the Oolitic period.

If our interest has been excited to a remarkable degree by the foreign appearance of the vegetation, with what surprise shall we view the forms of animate beings, which in this period peopled the earth and sea! With the exception of some great insects, such as the *Aeschna longiolata,* they consisted chiefly of amphibious animals. The gigantic remains of the *Ichthyosaurus platyodon* produce an extraordinary impression, whether they be seen as skeletons or almost in their natural state, like the amphibia of our own time, landed on sand-banks, and covered with slime and sea-weed. We observe, higher in the bay, another amphibious creature with fins, the long-necked *Plesiosaurus,* the *Plesiosaurus dolichodeiros,*

and in the air the singular *Pterodactylus,* whose fantastic form realizes the monstrous being that man's imagination has created under the name of the Flying Dragon.

Franz Unger, *Primitive World* (1851).

Text 41

Small, damp islands, covered with forests inhabited by the greatest and most terrible monsters of the ancient world: such are the scenes which this formation offers to the artist, judging from scientific researches already made [fig. 51]. An atmosphere filled with humid vapours and exhalations of carbonic acid was as favourable to this prodigious propagation of the amphibious races, as to the development of Ferns, Cycadeae, Coniferae, and of some Monocotyledons.

Agreeably to these data of science, we transport our reader into the twilight of a forest whose mysterious silence is interrupted only by the uniform dashing of a cascade, and the hissing and roaring of the monsters who dispute the sovereignty of the solitary regions. A warm and vaporous atmosphere covers with a thin veil all the objects at a little distance from the beholder. The straight-stemmed Tree-ferns placed on the naked Portland stone rocks, and Cycadeae, such as the *Zamiostrobus crassus,* and the *Pterophyllum Humboldtianum,* rise above the thickets which belong to the *Pecopteris,* the *Alethopteris* and the *Sphenopteris:* the largest and most beautiful of these kinds is the *Neuropteris Huttoni.* To the plants already named, we may add a tree which offers a very picturesque specimen of the great family of the Monocotyledons. It is the *Clathraria Lyellii,* little known until this period. The base of its trunk is encircled by the pretty foliage of the *Pterophyllum Schaumburgense;* and other tender sorts of Cycadeae, besides Ferns, cover the rocks or climb round the trunks of trees, to which they impart new vigour. The melancholy aspect of this forest is augmented by the strange nature of its inhabitants, among

which the gigantic bony-crested *Iguanodon* and the
monstrous *Hylaeosaurus* hold the prominent places.
Unfortunately, since only a few bones, teeth, and
fragments of the jaws of these animals have been
discovered, it has been necessary to trust to imagination for the greater part of the work of their
restoration, but we shall gradually have less and less
occasion to have recourse to its aid; for as new discoveries are made, less difficulty will be found in
divining approximately or ascertaining accurately
the physiognomy of these strange beings.

Franz Unger, *Primitive World* (1851).

Figure 51. "The Wealden [Lower
Cretaceous] Period," from Franz
Unger's *Primitive World* (1851).

119

ual on the right is strikingly similar in pose to Richardson's; the grouping of three at war is equally reminiscent of Martin's drawing. But Kuwasseg's rendering of the reptiles is more lifelike than Richardson's and more accurate than Martin's. Above all, it is ecologically more convincing: the three monsters are not tearing at each other cannibalistically, as in Martin's apocalyptic scene, but appear to be having a war of nerves over a potential mate. Unger, to his credit, notes that in any case these reptiles have been reconstructed from very sparse remains, and that future discoveries may modify their forms.[7]

The tenth scene (fig. 52; text 42) shows the period of the Chalk, the distinctive white limestone that is one of the most widespread formations in Europe. Its evidently marine origin faces Unger and Kuwasseg with the same pictorial problem as the Muschelkalk (fig. 48). The plant life clings precariously to some coastal rocks: although there are still tree-ferns and cycads, these are now joined by conifers, palms, and even the first dicotyledons, so that the vegetation is coming closer to its modern form. The marine life, on the other hand, although known from abundant Chalk fossils, is depicted only by a handful of mollusc shells thrown up on the shore. The vertebrates are represented only by the first true bird. In place of the desolate moonlight of the Muschelkalk scene, Unger here perhaps suggests his distaste for a primarily marine scene by having Kuwasseg portray it lashed by a storm.

The eleventh scene (fig. 53; text 43) moves the series into the early part of the Tertiary, the period of Cuvier's Montmartre mammals, or more specifically, of the coarse marine limestone (*calcaire grossier*) that is widespread around Paris. Now at last Unger's world has extensive areas of land as well as oceans, though the climate is still at least subtropical. The vegetation is fully modern, with conifers and palms, but above all dicotyledonous trees. It is indeed so modern that Unger judges the world almost

Text 42

The tempest which had agitated the immense cretaceous sea is now nearly over; in the distance only, thunder and rain still prevail; lightnings, at each flash less vivid, illumine at intervals the dark clouds that the rays of the setting sun, which gild the wall-formed rocks of chalk, have failed to disperse. Here [fig. 52] is represented a great gulf whose waters flow between the rocks of the small islands, as the appearance of the Swiss Jura valleys shows us now. A scanty vegetation has established itself on the more humid and less damaged banks, or has entrusted its mangled remains to the sea, which has hidden them as memorials of that period in the sandy stratum then in the process of formation (Quadersandstein). The Cycadeae and the Ferns are fewer in number: in their stead are seen new Palm-trees, Conifers, and, more remarkable, the earliest forms of the *Dicotyledons*. Let us examine them more minutely. The trees which present the most imposing appearance in this landscape are a group rising from those calcareous rocks whose summit the highest waves of the sea cannot reach. They belong to the genus *Credneria*, whose position in the vegetable world has not yet been assigned. The great leaves of these trees, which have three principal, crossed by numerous smaller veins, are the distinctive character of a genus in which difference of size alone constitutes diversity of kind. We have before us that which bore the largest and most remarkable leaves, the *Credneria subtriloba*. A second Dicotyledon, more to the right of the foreground, resembled the Willow in foliage; this is the *Salicites Petzeldianus*. In the centre rises a straight-stemmed Palm-tree, bearing leaves similar to those of our *Chamaerops*, thence called the *Flabellaria chamaeropifolia*: a trunk which has been torn up by the storm, and belongs to the same family, is visible in the background.

Finally, we perceive on the opposite coast some little slender and elegant trees, bearing simply a crown of leaves: they are arborescent ferns which still existed in the Cretaceous period (*Protopteris*

Figure 52. "The Period of the Chalk
[Upper Cretaceous]," from Franz
Unger's *Primitive World* (1851).

fit for human habitation, which previous periods have not been. Turtles and a wader in the foreground, and albatrosslike birds and a herd of palaeotherium in the background, represent the animal world. As Unger himself comments, the scene is tranquil and Arcadian.

The twelfth scene (fig. 54; text 44) represents a later part of the Tertiary, during which the important deposits of "Brown Coal," or lignite, were forming in what is now central Europe, where the scene is explicitly located. The swampy ground in the foreground suggests the accumulation of peat that will subsequently be turned into lignite; while the volcano in the background is based on the volcanic rocks of the period. The climate is now warm-temperate, like that of parts of the southern United States and Mexico at the present day; the vegetation is fully modern, with many familiar trees. Unger notes somewhat defensively that his scenes do not portray the animal world as fully as its importance warrants; but he does show some amphibians in the pool in the foreground, mastodons further back, and a cranelike bird in the air above. Although this was roughly the period of Kaup's dinotherium (fig. 30), Unger rather surprisingly fails to make use of that spectacular and well-known fossil.

The thirteenth and penultimate scene (fig. 55; text 45) represents the period that the Swiss zoologist Louis Agassiz (1807–73) had recently termed the Ice Age.[8] Like many other scientists, however, Unger adopted only a moderate version of the "glacial theory." Life on earth has not been virtually wiped out by an almost global sheet of ice, as Agassiz claimed; the climate has simply cooled further, to the point at which the elevation of a range of high mountains is enough to create glaciers on its flanks. This scene, like the previous one, is located quite specifically, this time on the northern edge of the Alps. In climatic though not scenic terms, Unger also likens it to the

continued on page 128

Singeri); and on the opposite side of the drawing is a little forest of Coniferae, which may be compared to trees resembling the *Cunninghamia* or *Damara* (*Cunninghamites oxycedrus, Damarites albens*).

The swollen waves caused by the storm have bathed the banks, and there deposited the remains of trees, of leaves, and of branches, besides seaweeds and testacea. Among the latter we observe chiefly the *Ammonites rhodomagensis*, the *Ostrea carinata*, the *Rostellaria Pes-Pelicani*; and farther from the sea, almost submerged, two shell-fish, one resembling the *Pecten quadricostatus*, the other, the *Cardium Hillanum*; lastly, the great *Inoceramus mytiloides*, which inhabited the rocks between the sea-grass and weeds. We notice here the first tenant of the air, the *Cimoliornis*, similar to the Albatros. The little *Protornis*, a bird of the size of a lark, was too small to be represented here.

Franz Unger, *Primitive World* (1851).

Text 43

Dry land has now gained in extent, and several islands unite to form one great tract. This Plate [fig. 53] transports us to the interior of one of the continents of the Eocene period. But a short time before, the waters overspread nearly the whole earth, leaving little room for vegetation; yet here we see a luxuriance of growth and a diversity of form which might serve as a type of our tropical landscapes. We have before us a mountain gorge, through which winds slowly a stream whose waters flow from an immense savannah. The chalky rocks, beaten by the tempest, are covered with plants, bushes, and trees of all kinds, in which we already perceive the promise of the world of man's future habitation. In the centre of the Plate we observe great trees with lobate leaves, and shrubs of delicate foliage: the former are of the beautiful family of the *Malvaceae*, the latter are Leguminous plants. Amongst the climbing plants which have spirally ascended the trunks of trees, will be observed the fossil species of *Cupanoi-*

Figure 53. "The Eocene Period
(Period of the Parisian *Calcaire
Grossier*)," from Franz Unger's
Primitive World (1851).

des, as well as the *Cucumites variabilis.*

To the left of the foreground are some beautiful Coniferae of the Cypress order (*Cupressites*); to the right, a Palm-tree, which may appropriately represent the *Palmacites echinatus.* The tall straight trunks of many Palm-trees adorn the landscape: they answer to the fossil remains of the *Flabellaria parisiensis,* and to some species of *Burtinia.* Lastly, the surface of the peaceful waters of the stream is embellished by several plants of the *Trapa* and of the *Potamogeton* families, and by the magnificent flowers and circular leaves of the *Nymphaea Arethusae,* Under the shadow of these plants, Turtles are reposing (*Trionyx*). Nothing interrupts the arcadian tranquillity of this region: neither the great *Haliaëtos* which hovers in the air, nor the *Tantalus* which rests on the riverbank; nor even the great herds of *Palaeotherium* which have retired to the prairies of this humid valley, and further animate this scene of primitive life.

Franz Unger, *Primitive World* (1851).

Text 44

The tropical scenes of the preceding period have now given place to those of the temperate zone: the great Palms and herbivorous Pachydermata, however, that we see here [fig. 54], constitute a similarity between this and the former view; though the mountains and valleys, the forests and plains, with their inhabitants, appear less strange, and strike us as half-familiar. Indeed, the basaltic domes of the Mittelgebirge in the distance resemble closely in their form and action our own volcanos: they may even serve as a representation of the chain of mountains which traverse the middle of Europe; since it was exactly that part of the globe which witnessed this phase of terrestrial development. If the mountains and the enormous marshy turf-beds have preserved their exact character to the present time, it must be more or less the same with the vegetable and animal world which animated this great stage. Favoured by soft airs, a bright sky, and abundant waters, the whole surface of the continent, not yet intersected by chains of high mountains, has gradually been clothed with a more luxuriant vegetation than any since that of the coal formation. As now, on the elevated plains, and the great valleys of our rivers, turf-beds, overshadowed by forest trees, then also accumulated an enormous quantity of vegetable substances, which form the greater part of the strata of this period, and which we use under the names Lignite and Coal. In this Plate we see, beyond the trees, a vast plain surrounding a lake formed by a great river. No thick vegetation intercepts the view, except a prairie of tufted trees that the river has flooded. In the foreground are stagnant waters bordered by reeds (*Culmites anomalus*); to the left is a picturesque entrance to a forest of trees grey with age, whose branches droop towards the ground. Like the angel of peace watching over a death-bed, the verdant species of *Smilacites* cling embracingly to these ancient trees, imparting to them new life. It is easy to recognize in the appearance of the trees, the various kinds of Maples, Chestnuts, Pollards, and Alders; and in the clusters of foliage which crown their summits, the *Phaenicites* and the *Flabellaria,* then so numerous. If we compare together the peculiarities of the flora of the Miocene and actual periods, we shall be struck with the resemblance which exists between the growth which distinguished the age we are contemplating, and that of the southern part of North America and Mexico. A similar, though less striking resemblance is traceable also in the animals, both small and large. It is, we trust, understood, that in our landscapes the animal world does not always preserve its comparative importance; the character of this period is, however, tolerably well represented in the gigantic Salamander (*Andrias Scheuchzeri*) which hides among the reeds of the marshes (the *Palaeophryne Gesneri*), in the *Dorcatherium Naui,* who with swift pace avoids the Mastodon that he sees among the thickets, and in the stork-like birds which animate the distance.

Franz Unger, *Primitive World* (1851).

Figure 54. "The Miocene Period
(Brown Coal Period)," from Franz
Unger's *Primitive World* (1851).

CHAPTER FOUR

Text 45

The scenes of a mild and propitious climate, which during the tertiary formations were gradually transformed into tropical and sub-tropical landscapes, have suddenly disappeared to give place to severer aspects. They are, indeed, to be found even now on the globe, but only under a vertical sun; the other regions of our planet have an altogether dissimilar character. Difference of climate necessarily infers variety of vegetation. While our preceding Plates might indifferently apply to any portion of the earth's surface, this, on the contrary, presents to us the aspect of a limited zone, far distant from the warmer climates. There still reigns a vegetation sufficiently luxuriant to nourish the great number of animals which peopled this region. The Plate before us [fig. 55] represents a portion of the interior of Europe to the north of the Alps, whose rugged heights have just risen; but it might also afford a not unfaithful picture of some appearances of the mouth of the Obi or the Lena. The chief charm belonging to scenes of this period is in the contrasts which strike us, whether we contemplate the air, the earth, the water, or the animal and vegetable worlds. It awakens in us, by its harmonious and powerful influence, the twofold admiration of the grace and of the might of nature. To produce this effect, the mountains, higher than any we have yet seen, doubtless contribute greatly.

It is a new world which has superseded the old; and which claims superiority as well by its extent as by the chaos of its varied forms. At this period of the prodigious extension of dry land, the general distribution of heat must have undergone considerable changes, and thus induced an entirely different order of appearances. The most important circumstance of this kind, as well as that which has most influence on the character of the landscape, is the first transformation of the waters into solid masses. During this change, which was effected on the summits of the highest mountains, so great a mass of atmospheric vapour was soon absorbed, that the ice,

much enlarged, descended to the valleys, forming glaciers, of which we see one to the left of the Plate. It has fallen with violence from the place of its origin, and lies between the rocks it has injured by the force of its descent. In other places, ice-walls fortify the valleys, and arrest the course of the waters, thus forming lakes both small and large; which, after a time, burst their enclosures, inundating the plains and deep valleys; and filling the ravines with great wrecks, and the clefts and subterraneous caves with diluvial slime. Coevally with this confusion was established a vigorous vegetation, consisting of trees proper to the north; forests of Conifers, of Oaks, and of Beeches, inhabited by numerous herds of animals. Of these, this landscape gives us some idea, in presenting to our notice, on one side an immense troop of *Bos priscus*, on the other a ferocious animal which dwelt in caverns, and is thence called the Cave Bear (*Ursus spelaeus*). We see it devouring the remains of a being much stronger than itself, whose great weight rendered it, however, inferior in combat, the Mammoth, or *Elephas primigenius*. Some birds of the Vulture kind seem anxious to share the monster's feast, but his ferocity is effectual in protecting his food by keeping the birds at a distance.

Franz Unger, *Primitive World* (1851).

Figure 55. "The Period of the Diluvium [Pleistocene]," from Franz Unger's *Primitive World* (1851).

tundra near the river Lena in northern Siberia, where the well-known frozen carcasses of woolly mammoths had been found.

Like most geologists at this time, Unger assumes that the elevation of the Alps, and hence also the formation of glaciers, have been geologically sudden events; but this does not materially affect his reconstruction. Most significantly, there is no question of any extensive—still less, worldwide—Deluge, even though Unger uses the term "Diluvium" as the title of the scene. Among German-speaking geologists, "Diluvium" had long become merely a technical term for the peculiar deposits earlier attributed (for example, by Buckland) to some kind of Deluge event. The term continued in use, even when those deposits came to be regarded as glacial in origin. Unger, like others, explains them in terms of sudden floods of purely local character, caused by the bursting of ice-dammed glacial lakes.

In correlation with a subarctic or at least Alpine climate, the woods are now of conifers, oaks, and beeches; and a herd of aurochs (the extinct wild bison) grazes at the edge of a glacial lake. But the scene is dominated by the cave bears outside their den, feasting on the remains of a mammoth. Once again, the ecological coherence of the scene is impressive. But equally striking is the fact that this is still a prehuman world. As Unger composed his text, the question of "the antiquity of man" was being actively debated elsewhere in Europe; but the general opinion was that there were no well-authenticated human fossils, nor any other reliable sign of human existence, from any deposits that could confidently be assigned to the glacial period.[9]

The fourteenth and final scene (fig. 56; text 46) purports to be of "the present world." But that title is a curious misnomer: in fact the scene is of the first human beings, who have at long last arrived on the world's stage.

Text 46

The loveliest day of creation dawns at length. In a pure and cloudless sky appears the day star, shedding its beams on that earth, which, after so many shocks and revolutions, has reached at last a state of repose [fig. 56]. The heavy, stifling atmosphere of a storm-clouded sky has given place to the sweet fresh air of spring; the earth has ceased to emit the noxious exhalations which formerly issued from it; no destructive revolution menaces our now tranquil planet. To the dry land, the sea-coast, the valleys, and the mountains, have been prescribed their precise and enduring limits; finally, peace is established between all the antagonistic powers which had so long contended, and which now seem to be reconciled in order to witness together the last and climacteric act of creation. For many ages creative power had been exercised on the production of numerous forms of plants and animals, always advancing from the simple to the complex, from masses roughly formed to nobler beings. Thousands of creatures were endued with life, all infinitely inferior to the work which was finally accomplished. Man appeared, the noblest creation of an omnipotent Master, whose will it was to vivify in him the thought of the universe.

Thus we see him appear among the most diversified existences; and of him alone may it be said for the first time, "The word was made flesh."

"In truth there was no necessity for the marvellous sowing of dragon's teeth to call him into being, for the germ of his life existed from the beginning of time, and awaited for its development only the coming of the appointed hour. Delighted, he contemplates himself as he rises from nature's slumber, and in gazing on the surrounding beauty, he understands the end of his existence."

It is at this moment that he is depicted in our landscape. All nature seems ready to obey him. No historian has yet attained a knowledge of the exact spot of man's origin; thus instead of copying faith-

Figure 56. "The Period of the
Present World," from Franz Unger's
Primitive World (1851).

fully the appearance of his paradise, we are obliged
to invent it. His birthplace, adorned by all the charms
of scenery that the serenest sky and the most fertiliz-
ing climate can produce, is here idealized. The Bro-
melia, the useful Banana, the gracefully formed
Palm, offer their nourishing fruits to the new comer,
who understood not the secrets of his domain till,
demanding from the earth new aliments, other
dwellings, and other modes of existence, he commenced a struggle with nature and with himself,—
the beginning of his eventful history.

Franz Unger, *Primitive World* (1851).

It is of the "present" only in the sense that it is *human*, and thereby contrasts with the *prehuman* "primitive world" portrayed in the first thirteen scenes. It also forms the culmination of the immeasurably long periods of deep time, and varied periods of deep history, represented in those scenes. Pictorially, Unger's world is dominated to the last by plants, most notably by the tree placed prominently in the center of the picture. But textually it is now dominated by the human presence, which evokes some of his most purple prose. For Unger, as for so many of his contemporaries, the entire history of life is a long story of the gradual preparation of the world for the coming of humanity. It is not for nothing that, quoting (but misapplying) the Fourth Gospel, Unger makes his Adam-figure the *logos* of the world, the Word that has at last been made flesh.

However, Unger has to admit that he has absolutely no evidence on which to base this pictorial reconstruction, and is therefore "obliged to invent it." In reality, Kuwásseg had plenty of precedents in the long artistic tradition of portraying the Garden of Eden and its secular equivalent, Arcadia. This is indeed an Edenic scene, with a Tree of Life at its center. The human family in primal nakedness does not quite correspond to its original, but the parallel is clear. The human beings are unmistakably white, European, and civilized. Their essential *difference* from the whole of the natural world seems to be accentuated by the sharply outlined style in which they are drawn.

Equally, this is an Arcadian scene, in which nothing is lacking to human needs. The man holds only a staff, not a weapon of any kind. Only some playful horses in the background suggest a wilder Nature, and even they seem to invite domestication to human use. The only hint of future problems, of a fall from grace, is in Unger's final comment: now, at the end of the long and varied history of the "primitive world," comes the start of man's "struggle with Nature and with himself, the beginning of his eventful [human] history."

Unger, like all his scientific colleagues, believed that this progressive development of the earth and its life had been spread over periods of time that were of literally unimaginable magnitude. But that does not alter the close parallel between this sequence of scenes from deep time, based on some of the best science of the mid nineteenth century, and the traditional sequences of scenes of the "days" of Creation, for example, those of Scheuchzer (figs. 1–6). The contrast in their respective time scales is almost irrelevant; what is striking is the similarity, indeed virtual identity, of their imaginative evocations of a world that has changed from the alien into the familiar, from the nonhuman into the human.

Unger and Kuwasseg's publication is of outstanding importance in the history of scenes from deep time, but its initial impact was limited. As already mentioned, the text was published not only in Unger's native German but also in French, which would have made it linguistically accessible to scientists of any standing throughout the world. But in practice, its *de luxe* format and a high price to match seem to have restricted the range of those who became aware of its existence. The list of the forty-six subscribers who insured the publisher against loss gives, of course, only a minimal indication of the print run, but it does suggest that the number of copies printed may have been quite small.[10]

However, even if that was the case, the list does indicate the kind of distribution that the work received immediately. The subscribers included, for example, university scientists at Bern, Dresden, Giessen, Tübingen, and Vienna in the German-speaking world; and Edinburgh, Geneva, Harvard, Leiden, and Padua outside it. Copies went to scientific academies in Vienna and

St. Petersburg, and to booksellers in London and Munich. The work also graced the libraries of royal, aristocratic, and ecclesiastical patrons, headed by the King of Saxony. The French were curiously absent from the list, and English-speakers were represented only by the Scot David Brewster and the American Asa Gray; but German-speaking central Europe was well covered. All in all, the work evidently got into at least a select number of well-qualified hands.

The great sequence of scenes that Kuwasseg drew for Unger has deserved a chapter to itself, because, apart from the minor exceptions of Trimmer's and Reynolds's designs, it was the first use of scenes from deep time to suggest the temporal development of the natural world. The Edenic overtones of its culminating final scene, with its overtly biblical allusions, suggest that it is not fanciful to see in Unger's work the definitive assimilation of the tradition of biblical illustration into the newer genre. As already suggested, the difference between the vast (though unquantifiable) time scale that Unger adopted from the geologists of his time and the very short time scale of the older tradition (as exemplified by Scheuchzer's work) is far less important historically than their common vision of a purposeful development from the earliest epochs toward the human world.

By the time Unger's work was published, that word "development" was commonly used as a synonym for "transformism," or what modern scientists term "evolution." So it is important to note that a sequence of scenes such as Unger's was essentially neutral with respect to any and all of the many evolutionary speculations that were "in the air" at this period. Unger himself, like many other naturalists, apparently believed that some kind of natural process had produced the diversification of the living world in the course of earth history, though he would probably not have welcomed the materialistic implications of the particular theory that the English naturalist Charles Darwin (1809–82) was formulating privately around this time.

One other feature also makes Unger's work an important landmark in the history of scenes from deep time. The earliest true scene, De la Beche's *Duria antiquior,* was drawn by a scientist who happened to be a good amateur artist. By contrast, later authors usually recruited a professional artist, usually a landscape painter or book illustrator, to do the drawing to their instructions. But few—Trimmer and, of course, Mantell are exceptions—gave their artists much explicit credit. In Unger's work, however, the artist Kuwasseg is for the first time given great prominence and much of the credit for the success of the scenes. The normally "invisible" craftsman or technician in scientific work becomes, here at least, highly visible.[11]

5

Domesticating the Monsters

Kuwasseg's magnificent lithographs, executed under Unger's scientific direction, represented a striking vision of a differentiated and ever-changing "primitive world." But owing to the limited distribution of their work, many of the scenes published in the following years still showed little acknowledgment of that concept. Indeed, most popular books on geology continued to display the single scene, by now almost obligatory, centered on the Liassic reptiles, which purported to represent the whole of pre-human nature. A couple of examples are sufficient to illustrate the genre.

The Wonder of the Primitive World (*Die Wunder der Urwelt*, 1855), by W. F. A. Zimmermann—a pseudonym for the German writer W. F. Volliner—was a highly popular work that went through several editions in both German and French. Its frontispiece, predictably entitled "The Primitive World," displayed the usual battling reptiles against a background of the usual sample of Jurassic plants (fig. 57). Some of the reptiles seem to have been borrowed from Martin's scene for Richardson (fig. 37), while the pterodactyles are modified—or corrupted—into batlike forms. Elsewhere in the book there are several reconstructed skeletons and body outlines of animals such as the mammoth and the "Irish elk," which by now were almost as *de rigueur* as the Liassic reptiles, but the frontispiece is the only true landscape scene from deep time. The general message conveyed by the book, which was intended to be highlighted by its frontispiece, was the now stereotyped "wonder" of the immense scale of the earth's prehuman history (text 47).[1]

More interesting in design is the small frontispiece of an otherwise unoriginal compilation of Victorian natural theology entitled *Theology in Science* (1860), which was intended "for the use of schools and of private readers." The author was Ebenezer Cobham Brewer (1810–97), a clerical don at Cambridge and a prolific author of textbooks and reference works, who is best known for his later and eponymous *Dictionary of Phrase and Fable* (1870). It is characteristic of the period that, in a book covering a very wide range of sciences, Brewer chose a geological picture for his frontispiece. It ingeniously combines a scene from deep time with a section of the strata from which the scene is inferred, separated by a conventional barrier of the nonstratified rocks basalt and granite (fig. 58). As usual, the scene is based mainly, though probably indirectly, on De la Beche's Liassic reptiles.[2]

The same year, a scene in another popular English book broke this standard mold by suggesting a sequence in deep time. Its design recalls Trimmer's layered pile of discrete scenes (fig. 39) rather than Reynolds's continuous panorama (fig. 42), but it was intended to illustrate a theme far less prosaic than Trimmer's practical geology and far less conventional than Reynolds's anodyne natural theology. It was a large fold-out plate that decorated a book entitled *Pre-Adamite Man* (1860), by Isabella Duncan. The author is now obscure, but in its time the book was evidently not: it had reached a fourth edition within a couple of years.

Text 47

But geology—what is its basis? *The sand and the stone that we tread underfoot.* And yet, *one has been able to assemble the archives of the primitive world*, and to extract from this immense treasure the positive and distinct history of the different epochs of the globe, of the generations of plants and animals, the age of which reaches back to such an antiquity, that by comparison the age of the human species is a mere zero, and that a thousand times that age would still be a trifle in relation to the antiquity of the world.

W. F. A. Zimmermann, *Wonder of the Primitive World* (1855).

Figure 57. "The Primitive World":
the frontispiece of W. F. Volliner's
Wonder of the Primitive World (1855),
published under his pseudonym of
W. F. A. Zimmermann.

The lowest and oldest layer of Duncan's design (fig. 59; text 48) presents the usual reptiles (and Owen's amphibian labyrinthodon) from all the Secondary formations. The second scene likewise lumps together all the Tertiary mammals, from Cuvier's Eocene forms to the much younger mammoth and "Irish elk." The boundary between these two "platforms," as Duncan calls them, is relatively indistinct—the lizardlike iguanodon pokes its snout out of the Secondary into the Tertiary!—so that the effect at that point has the continuity of Reynolds's panorama. But the third scene explicitly reflects Duncan's adoption of Agassiz's theory of the Ice Age in its fullest form; it portrays a lifeless ice-bound world that forms an absolute barrier between the ancient and modern worlds. The fourth scene, at the top, depicts that present world, with a selection of living animals; the pyramids of ancient Egypt in the far distance allude to an early human presence.

The design as a whole gives quite an effective impression of at least a few successive phases in the history of life; it clearly portrays "the story of our old planet and its inhabitants"—the subtitle of the book—as a story with successive chapters. What could hardly be guessed from the illustration alone, however, is that all this geological history is deployed in the service of a revival of the seventeenth-century theory of Pre-Adamites, as first popularized in Isaac La Peyrère's notorious book *Men before Adam* (*Prae-Adamitae,* 1655).[3] Duncan uses the total break of the Ice Age to separate the world of modern men or "Adamites,"—back beyond the ancient Egyptians and their pyramids to Adam himself—from the world of the "Pre-Adamites," who had ruled the ancient world revealed by geological research. These early men, she argues, had left traces of their presence in the form of the flint axes that were being found with the bones of extinct mammals in ancient river gravels; but since there was

Text 48
EXPLANATION OF THE PLATE

As many of my readers may not be familiar with the restorations here delineated, I have thought it may be acceptable to devote a few pages to an explanation of these singular productions of former times, the representations of which are here introduced. Our artist has drawn the principal animals, constituting the fauna of three prolific periods, enabling my readers to contrast the several characters of them with one another; the secondary on the lowest platform; the tertiary on that which occupies the middle, and our own at the top. No attempt is made to distinguish the animals which belonged alike to both of the earlier epochs, though a certain degree of continuity is indicated, and it will be noticed that between the tertiaries and our own times he has tried to exhibit to the eye that singular crisis in the history of our planet where a destructive glacial influence is supposed to have invaded the earth and to have put a sudden end to life, both in the vegetable and in the animal kingdom—a crisis which occupies a somewhat conspicuous place in our pre-Adamite theory.

Isabella Duncan, *Pre-Adamite Man* (1860).

Figure 58. "The Rocks and Ante-diluvian Animals": the frontispiece of Ebenezer Brewer's *Theology in Science* (1860).

THE ROCKS AND ANTEDILUVIAN ANIMALS.

Figure 59. A panorama of the modern "Adamitic" world (top), separated by the lifeless Ice Age from the "Pre-Adamitic" world of the Tertiary mammals and Secondary reptiles (below): lithograph by W. R. Woods, from Isabella Duncan, *Pre-Adamite Man* (1860).

139

no trace of their skeletons, Duncan inferred that their bodies must have been transmuted and transported to some other region of the universe.

Such a theory may now seem bizarre, but it belongs—like Thomas Hawkins's ideas—to a flourishing Anglo-American subculture of biblically based cosmological speculation, often with powerful social and racial implications. In the present context, however, the importance of Duncan's book is that it shows not only how widely the genre of scenes from deep time had spread by about 1860, but also how versatile such scenes could be, in the service of diverse interpretations of the history of life. The scenes described so far in this chapter were, not surprisingly, derivative in content; but they certainly introduced an ever-widening public to the strange creatures of the deep past.

That penetration of "antediluvian monsters" into the cultural consciousness of the nineteenth century was due above all, however, to the impact of the first major display of the kind that still draws crowds to our modern museums: a display of full-scale three-dimensional reconstructions. This was in fact opened to the public several years before Duncan's scene was published, and her artist was able to borrow extensively from it. The narrative thread followed in this book therefore needs to loop backward at this point to the early 1850s.

When the Great Exhibition of 1851 in London—the first major international event of its kind—finally closed, its huge and spectacular steel and glass building in Hyde Park was dismantled and reerected in the suburb of Sydenham, where it became the "Crystal Palace." It was to be the centerpiece of a permanent exhibition of the arts and sciences, set in extensive grounds and developed by commercial enterprise. The Crystal Palace was easily accessible by inexpensive suburban train from the crowded housing of central London, and it quickly became a

Text 49

In the first week of September, 1852, I entered upon my engagement to make [a] mastodon, or any other models of the extinct animals that I might find most practicable; such was the tenour of my undertaking, and being deeply impressed with its important and perfectly novel character, without precedent of any kind, I found it necessary earnestly and carefully to study the elaborate descriptions of Baron Cuvier, but more particularly the learned writings of our British Cuvier, Professor Owen. Here I found abundant material collected together, stores of knowledge, from years of labour, impressing me still more with the grave importance of attempting to present to the eye of the world at large a representation of the complete and living forms of those beings, the minutest portion of whose bones had occupied the study and research of our most profound philosophers; by careful study of their works, I qualified myself to make preliminary drawings, with careful measurement of the fossil bones in our Museum of the College of Surgeons, British Museum, and Geological Society; thus prepared I made my sketch-models to scale, either a 6th or 12th of the natural size, designing such attitudes as my long acquaintance with the recent and living forms of the animal kingdom enabled me to adapt to the extinct species I was endeavouring to restore. These sketch models I submitted in all instances to the criticism of Professor Owen, who with his great knowledge and profound learning most liberally aided me in every difficulty. As in the first instance it was by the light of his writings that I was enabled to interpret the fossils that I examined and compared, so it was by his criticism that I found myself guided and improved, by his profound learning being brought to bear upon my exertions to realise the truth. His sanction and approbation obtained, I caused the clay model to be built of the natural size by measurement from the sketch-model, and when it approximated to the form, I with my own hand in all instances secured the anatomical details and the characteristics of its nature.

Some of these models contained 30 tons of clay, which had to be supported on four legs, as their natural history characteristics would not allow of my having recourse to any of the expedients for support allowed to sculptors in an ordinary case. I could have no trees, nor rocks, nor foliage to support these great bodies, which, to be natural, must be supported fairly on their four legs. In the instance of the Iguanodon [it] is not less than building a house on four columns, as the quantities of material of which the standing Iguanodon is composed, consist of 4 iron columns 9 feet long by 7 inches diameter, 600 bricks, 650 5-inch half-round drain tiles, 900 plain tiles, 38 casks of cement, 90 casks of broken stone, making a total of 640 bushels of artificial stone.

These, with 100 feet of iron hooping and 20 feet of cube inch bar, constitute the bones, sinews, and muscles of this large model, the largest of which there is any record of a casting being made.

I have only to add that my earnest anxiety to render my restorations truthful and trustworthy lessons has made me seek diligently for the truth and the reward of Professor Owen's sanction and approval; which I have been so fortunate as to obtain, and my next sincere wish is that, thus sanctioned, they may, in conjunction with the visual lessons in every department of art, so establish the efficiency and facilities of visual education as to prove one of many sources of profit to the shareholders of the Crystal Palace Company.

Waterhouse Hawkins, lecture to the Society of Arts, London (1854).

popular weekend excursion for a wide social range of London's vast and rapidly expanding population.[4]

Among the attractions planned for the grounds of the Crystal Palace was a series of life-sized reconstructions of some of the more spectacular fossil animals that geological research had revealed in the preceding decades. The Crystal Palace Company awarded this commission to the London sculptor and illustrator Benjamin Waterhouse Hawkins (1807–89) (not to be confused with the eccentric fossil collector Thomas Hawkins who figured earlier in this book). The artist was already familiar with scientific illustration: his were the drawings in the report on reptiles (1842–45) from the voyage of the *Beagle,* on which the young Charles Darwin had been unofficial naturalist. For his work at the Crystal Palace, Hawkins had the scientific help of the "British Cuvier," the anatomist Richard Owen. Their working partnership became as close as that between Unger and Kuwasseg. Owen expounded his own ideas about the original forms of the animals; Hawkins made scale models for Owen's approval and then prepared to expand them up to life size. In a large temporary workshop on the Crystal Palace site, using vast quantities of mundane materials, the extinct animals gradually took on huge and three-dimensional shape (fig. 60; text 49).

Of all the animals that Hawkins was reconstructing, the iguanodon was of special theoretical importance to Owen. Back in 1841, at the meeting in Plymouth of the British Association for the Advancement of Science, Owen had reviewed all the fossil reptiles from the Secondary formations and had erected a major new category, *Dinosauria,* for the iguanodon and some other forms.[5] The innovation was far more than merely taxonomic. On the basis of no more than some teeth and a few bones, Owen inferred that these reptiles had been almost mammalian, not only in anatomy but also in physiology. The boldness of this reconstruction was powered by theoretical, even ideological considerations. Owen was determined to turn the fossils into authoritative evidence against the rising tide of evolutionary speculation in the Lamarckian mode, which he regarded as deeply threatening to both science and society. That involved a dramatic transformation in the form of the iguanodon, from Mantell's huge crawling lizard (fig. 34) into a rhinoceros-like form standing upright on four massive legs. Only if this relatively ancient reptile had had a more "advanced" anatomy and physiology than living reptiles such as crocodiles—indeed, as advanced as living *mammals* such as elephants and rhinoceroses—could it refute the intrinsic progressiveness that Lamarckians claimed to see in the history of life.[6] It is therefore no wonder that under Owen's tutelage, Hawkins made the iguanodon the spectacular centerpiece of his workshop, which a newspaper artist duly portrayed as a stable of monsters (fig. 60).

The iguanodon was also the site of a celebrated dinner *inside* its reconstructed body (fig. 61; text 50). In the marquee that was erected around the model, the names of Cuvier, Buckland, Mantell, and Owen were all prominent. Owen, the only active survivor of this gallery of dinosaur heroes, presided at the head of the table; as the brains behind the whole project, he was seated appropriately inside the reptile's head. The occasion was not only excellent advance publicity for Hawkins's exhibit; it was also explicitly an opportunity for Owen and Hawkins to reinforce the scientific authority of their joint work. The placarded names of earlier heroes of dinosaur research tacitly recruited *their* authority for the reconstructions; Owen's speech allied the work even more explicitly with Cuvier, whose name, even two decades after his death, was still one to conjure with. The event showed how the genre of scenes from deep time still needed rhetorical

Text 50

On Saturday evening last (the last day of the year 1853), . . . Mr W. Hawkins, with the concurrence of the Directors [of the Crystal Palace Company], invited a number of his scientific friends and supporters to dine with him in the body of one of his largest models, called the Iguanodon, which occupies so conspicuous a place in our Illustration of last week [fig. 60]. In the mould of this colossal work of art—for as such it must deservedly rank very high—Mr. Hawkins conceived the idea of bringing together those great names whose high position in the science of palaeontology and geology would form the best guarantee for the severe truthfulness of his works. . . .

To carry out this extraordinary idea, cards were issued at the beginning of last week—and such cards! as startling as the invitation they bore: "Mr B. Waterhouse Hawkins solicits the honour of Professor ———'s company at dinner, *in the Iguanodon,* on the 31st of December, 1853, at four P.M." The incredible request was written on the wing of a Pterodactyle, spread before a most graphic etching of the Iguanodon, with his socially loaded stomach. . . . Mr Hawkins had one-and-twenty guests around him in the body of the Iguanodon on Saturday last: at the head of whom, most appropriately, and in the head of the gigantic animal, sat Professor Owen [fig. 61]. . . .

The dinner, which was luxurious and elegantly served, being ended, the usual routine of loyal toasts were duly given and responded to—allusion being gracefully made by Mr Francis Fuller, Managing Director, to the great interest evinced and approbation expressed by H.M. the Queen [Victoria] and H.R.H. the Prince [Albert], on their recent visit to the extraordinary works by which the company was surrounded.

Professor Owen then took occasion to explain, in his lucid and powerful manner, the means and care-

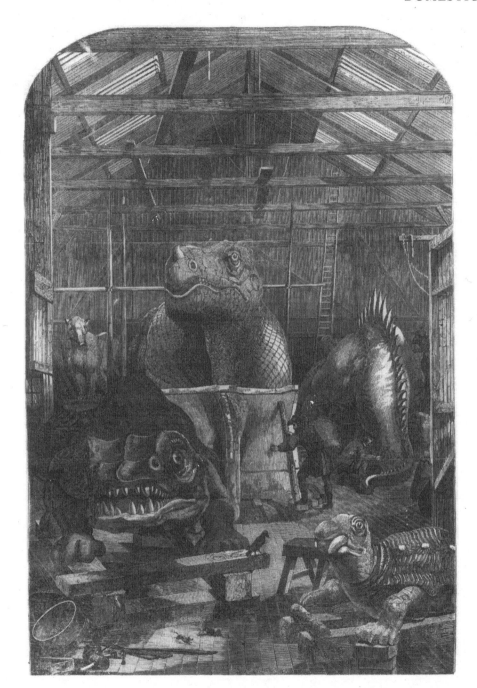

Figure 60. Waterhouse Hawkins's "Extinct Animals Model-Room, at the Crystal Palace, Sydenham." From the *Illustrated London News* (1853).

hard work if it was to be brought into the mainstream of respectable science.

However, while the models remained in Hawkins's workshop, they were still merely three-dimensional versions of the reconstructed bodies that Cuvier had first pioneered as two-dimensional drawings (figs. 15, 16). The full force of the lesson in the history of life that Owen and Hawkins had planned only became apparent when the models were taken out into the grounds of the Crystal Palace. There the wonders of modern science, in the form of a newly *material* scene from deep time, could be viewed against the backdrop of a building that embodied the equal wonders of modern technology (fig. 62; text 51).

The models were placed on an artificial island, no doubt to keep them safe from prying hands and clambering children. But whether intentionally or not, this also created an appropriate sense of distance between the "inhabitants of the ancient world," as Owen's threepenny guidebook called them, and the human spectators in the modern world. Whereas Unger's scenes had focused on the plant life of the "ancient world," this exhibit echoed most other earlier scenes by casting the animals in the main roles; the "appropriate vegetation" in which they were set was simply selected from horticultural nurseries. Still, the naturalistic illusion was to be heightened by artificial tides, which would alternately submerge and reveal the marine reptiles of the Liassic period.

The scene from deep time that Hawkins set on this island matched the modest graphic design that Emslie had made for Reynolds a few years earlier (fig. 42). For the island was not just a three-dimensional glimpse of "the country of the iguanodon," as Mantell had titled Martin's scene (fig. 35). It was intended to be a *temporal* panorama that stretched unbroken from Owen's labyrinthodon in the time of the "New Red Sandstone" or Trias, through

ful study by which Mr. Hawkins had prepared his models, and had attained his present truthful success; Professor Owen adding that it had been a source of great pleasure to him to aid so important an undertaking, by assisting with his instruction and direction a gentleman who possessed the rarely united capabilities of an anatomist, a naturalist, and a practical artist, with a docility and eagerness for the truth which ensured Mr. Hawkins's careful restorations the highest point of knowledge which had been attained up to the present period. The learned Professor then briefly commented upon the course of reasoning by which Cuvier, and other comparative anatomists, were enabled to build up the various animals of which but small remains were at first presented to their anxious study; but which, when afterwards increased, served to develop and confirm their confident conceptions—instancing the Megalosaurus, the Iguanodon, and Dinornis as striking examples.

. . . After several appropriate toasts, this agreeable party of philosophers returned to London by rail, evidently well pleased with the modern hospitality of the Iguanodon, whose ancient sides there is no reason to suppose had ever before been shaken with philosophic mirth.

Illustrated London News (1854).

NOTE. *Dinornis:* the moa, a large extinct flightless bird from New Zealand, which Owen himself had reconstructed from scanty fossil remains.

Text 51

Hitherto we have been accustomed to wonder at or study these monsters of the old world, either in pictures of a small size, illustrating the descriptions of writers on the subject; or in museums where a vast fragmentary, or an almost complete fossil skeleton, gave us a vague idea of these predecessors of the living family of nature. But at Sydenham we are not to be contented with either pictures or dry bones.

The Gardens and Park are sufficiently advanced

Figure 61. "Dinner in the Igua-
nodon Model, at the Crystal Palace,
Sydenham." From the
Illustrated London News (1854).

Figure 62. Waterhouse Hawkins's
models of extinct animals, in the
grounds of the Crystal Palace: a
vignette from Richard Owen's
guidebook (1854).

the period of the now familiar marine reptiles of the Lias, into the time of the iguanodon and the Wealden deposits, and as far as the period of the Chalk and its pterodactyles (fig. 63). Indeed, the map in Owen's guidebook suggests that the panorama was originally planned to continue, on a separate island, into the Tertiary and its mammals, including presumably the mastodon that Hawkins said he had first been commissioned to produce (text 52).[7] This unbroken sweep of deep time was matched, as in Reynolds's scheme, by a display of the succession of strata from which the temporal sequence was inferred; but here, in place of a conventional graphic column of formations, an artificial cliff behind the animals revealed a condensed replica of the strata.[8] (Hawkins's models survived the total destruction of the Crystal Palace by fire in 1936, and can still be seen in the public park at Sydenham.)[9]

What Emslie had drawn for Reynolds on a couple of broadsheets was now laid out by Hawkins, to Owen's instructions, on an island in the middle of pleasure gardens in a London suburb. Hawkins said he regarded the whole Crystal Palace enterprise, and his own exhibit in particular, as "one vast and combined experiment of visual education" that gave it "legitimate claims to the support of all civilised Europe."[10]

Certainly his work made a scene from deep time vividly accessible to a vast new public. The grounds of the Crystal Palace were opened by Queen Victoria in 1854 to a crowd of 40,000 people, brought from London in special excursion trains; and that was only the start of a period of intense popular interest in the exhibits. Specifically, the fame of Hawkins's display, as the first major three-dimensional reconstruction of its kind, spread quickly throughout Europe and North America. That it embodied Owen's subtle anti-Lamarckian message may not have been generally appreciated; nor that it

to enable us to form a tolerably correct impression of the general plan in which Sir Joseph Paxton has endeavoured to produce a fitting foreground to his Palace.

. . . One of these winding paths will lead . . . to a pool of about six acres, which will receive, by open and secret channels, the waters of the larger basins. On this tidal pool, at a convenient distance from the spectators, islands of irregular shapes will be placed, and covered with luxuriant vegetation. On one of these islands will be placed, in natural attitudes, and amid an appropriate vegetation, animals of the secondary, and others of the tertiary period; while opposite to each will be full-proportioned representations of the strata in which the remains of these vast beasts were found. To add to the illusion, the waters of the pool will rise and fall, partially submerging the amphibious inmates from three to eight feet alternately, during the playing of the great waters, after the manner of an actual tide. Thus, then we shall see, pausing among the rushes, the Iguanodon, or monstrous lizard, thirty feet high, and a hundred feet from snout to tip of tail. The Megatherium or monster Sloth, will appear in the act of climbing an antediluvian tree; huge Chelonians [turtles] are to bask upon its banks. The Plesiosaur, with its reptile form and bird-like neck, will wallow in the mud; while the Brobdignagian grandfather of turtles, gaping, shall frighten aldermen with ideas of retribution in its monstrous jaws.

And so the visitor, exhausted with the marvels of ancient art and modern commerce and ingenuity, within the Building . . . will be able to find fresh air amid all the stately luxury of Italian gardens outside . . . ; unless he prefers to turn aside and study, in strata and strange beasts, the history of that old world before which the trophies and gods of Assyrian kings are but as of yesterday.

Illustrated London News (1853).

Text 52
The diagram [fig. 63] shews those formations which
constitute the secondary epoch, or, if described in
ascending order, the commencement of that vertebrate existence which left unequivocal evidence of its
inhabiting the earth, by leaving the imprint of its
footmarks, which, at one time, was all we knew of
the extraordinary inhabitants of the New Red Sandstone, when it was called Chirotherium, from the
hand-like shape of the foot-marks, until the mighty
genius of Professor Owen placed the teeth and head

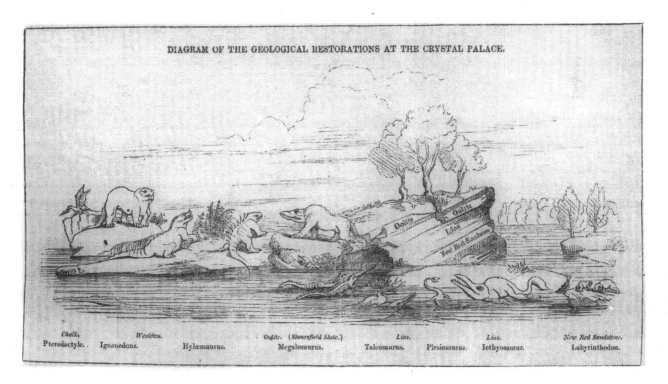

DIAGRAM OF THE GEOLOGICAL RESTORATIONS AT THE CRYSTAL PALACE.

Chalk.   *Wealden.*   *Oolite.* (*Stonesfield Slate.*)   *Lias.*   *Lias.*   *New Red Sandstone.*
Pterodactyle.   Iguanodons.   Hylæosaurus.   Megalosaurus.   Teleosaurus.   Plesiosaurus.   Icthyosaurus.   Labyrinthodon.

Figure 63. "Diagram of the Geological Restorations at the Crystal
Palace," drawn by Waterhouse
Hawkins to illustrate his lecture to
the Society of Arts in London
(1854). The models are arranged in
chronological order (right to left),
corresponding to the order of the
strata (in the background cliff face).

147

portrayed not just one scene but a whole temporal panorama. But its obvious parallel to the living exhibits at the Zoological Gardens a few miles away in Regent's Park must surely have brought the spectacular otherness of "the ancient world" vividly to the public imagination.[11]

Certainly that was the conception expressed by John Leech (1817–64), who at this time was one of the most popular and prolific cartoonists for the humorous magazine *Punch*. Just as Conybeare's cartoon had shown Buckland penetrating through time into a den of fossil hyenas (fig. 17), so Leech drew a Victorian paterfamilias walking a reluctant child through a crowd of alarming "antediluvian" monsters (fig. 64). The epistemic barrier symbolized by the water that lay between Hawkins's models and the visitors to Sydenham was here ignored. In imagination if not in reality, the Victorian public could believe themselves transported bodily into the deep past, and supposedly have their minds improved by the experience, while remaining safely within sight of a familiar English landscape (indicated by the distant church spire).

Another *Punch* cartoon complemented Leech's design by highlighting the nightmarish sensation that a tour of the "Antediluvian Department" evidently evoked, even in some of its adult visitors (fig. 65). After too good a dinner, the dreamer senses himself surrounded by these fearsome creatures, all too much alive, and mixed incongruously with the ethnographic displays of primitive Africans and ancient Egyptians that he had seen in the Crystal Palace itself.

Hawkins's display of monsters, however great its impact on the public imagination, was of course immobile. Its impact could only extend to those unable to visit the Crystal Palace if the models were translated into mobile forms. In this respect, the small-scale models that were soon marketed commercially—the forerunners of some of the most popular items in modern museum shops—

before us, with such indisputable characters as united them to the footmarks, and thus, by induction, the whole animal was presented to us.

Next, in ascending succession, we have the Tethyosaurus, Platyodon, Tenuirostris, and Communis, the Plesiosaurus Dolichodirus, as restored by Dean Conybeare, the Plesiosaurus Macrocephalus and Hawkinsii, the latter named by Professor Owen after Mr. Thomas Hawkins, who with great enthusiasm cleared it from its matrix of lias, and made the first great collection of fossils of the lias which were purchased by the trustees of the British Museum, where they are now, and form the most striking features of the national collection of fossils.

It next illustrates the upper portion of the lias, sometimes known as the alum shale, so well developed at Whitby, in which remains of the Teleosaurus have been so frequently found. This animal will be recognised by its near resemblance to the crocodile of the Ganges called Gavial, or Garrial, as it should be called: to the casual observer the principal difference consists in its greater size. The next formation above the lias is the oolite, of which at present that singular reptile, the Pterodactyle, represents the inhabitants, while the intermediate formation, called the Stonesfield slate, bears the great discovery of Buckland, the Megalosaurus, or the great lizard. This, the upward strata of the great oolite, brings us to the formation called the Wealden, which Professor Owen, in one of his elaborate descriptions of the British fossil reptiles, calls the metropolis of the Dinosaurian order, which I have here represented by the best known and most typical species, the Hylaesaurus or lizard of the mud, with its extraordinary dermal covering and long range of dorsal scutes, of which the bones were found by the late Dr. Mantell, whose persevering researches in Wealden formations first gave the idea to science of the former existence of the Iguanodon.

Waterhouse Hawkins, lecture to Society of Arts, London (1854).

A VISIT TO THE ANTEDILUVIAN REPTILES AT SYDENHAM—MASTER TOM STRONGLY OBJECTS TO HAVING HIS MIND IMPROVED.

Figure 64. A cartoon by John Leech for *Punch* (1855). This is a *middle-class* "Master" Tom who is having his mind improved by science; the improvement is not the anti-evolutionary message that the working class was considered to need, to keep it politically docile.

were probably important. But they were only models of individual animals, not of whole scenes from the deep past. To convey a fuller impression of the prehuman world, two-dimensional forms were essential.

In the years that followed the opening of the Crystal Palace exhibit, the kind of modest two-dimensional scene reviewed earlier in this chapter began to show the impact of Hawkins's reconstructions: many of the animals in Duncan's design (fig. 59), for example, are clearly modeled on Hawkins's exhibit. One such scene was drawn by Hawkins himself. Among the sources for his models had been the illustrations of skeletons in Buckland's Bridgewater Treatise (1836). It may have been the success of the new exhibit that led Buckland's son, the naturalist Francis Buckland (1826–80), to reissue that still popular work after his father's death in 1856. In any event, Hawkins repaid his implicit debt by contributing a new scene of the Liassic reptiles (fig. 66; text 53). The design was unoriginal and unremarkable, except for the newly monstrous appearance of the ichthyosaur; but it does show how Hawkins's skill as a sculptor was reflected in his style as a graphic artist. Thus Buckland's book, in this new edition (1858), came to include posthumously a bold scene from deep time, of the kind that the author in his lifetime had so conspicuously avoided, even in his popular work.

A more important outcome of the Crystal Palace exhibit was the incentive it seems to have given the London publisher Samuel Highley to issue an English edition (1855) of Unger's work. This was explicitly intended to complement Hawkins's extinct animals with similar views of the ancient world of plants. Highley's edition made Unger's text more accessible to the English-speaking world, which then as now was notorious for linguistic laziness. But in addition, by using the still quite novel technique of photography, it made Kuwasseg's litho-

Figure 65. "The Effects of a Hearty
Dinner after Visiting the Antediluvian
Department at the Crystal Palace":
a cartoon from *Punch* (1855).

graphs more widely known, although at a much reduced and less impressive size.[12]

There is no indication that Highley's edition had Unger's prior approval: in the absence of international copyright agreements, such book piracy was common. Its publication may well have been one factor in Unger's own decision to publish a new edition of his *Primitive World* in its original *de luxe* format, with text in German and French (1858). The work was almost unchanged, except that a new preface (text 54) explained that further research in the past decade justified adding two new scenes at the start of the series. These extended the sequence backward in time into those parts of the older Transition that Murchison had termed the Silurian and Devonian periods; as Unger put it, they formed a kind of prologue to the history of life previously portrayed.

The importance of Unger's new opening scene, of the Silurian (fig. 67; text 55), is that it was the first attempt to portray a period before the appearance of any terrestrial life at all. Following the broad geological interpretation embodied in his original sequence, this is a world of limitless ocean, without even a few islands. Its temporal proximity to the origin of life itself is suggested visually by Kuwasseg, with the rays of the rising sun bursting through onto this "watery desert"; and verbally by Unger, with references to the simultaneous creation of animals and plants, and to the divine breath still hovering over the waters. Since an underwater scene was inconceivable—De la Beche's *Duria antiquior* (fig. 19) was either still unknown to them or else unacceptable as a model—the marine life of this early period is depicted in the now traditional manner as a set of shells and seaweeds cast up on the shore.

With Kuwasseg's second new lithograph, of the Devonian (fig. 68; text 56), Unger is able to turn with evident relief to a terrestrial and botanical scene. The first few

Text 53

By the kindness of my friend, Mr. Waterhouse Hawkins, who, after a continuous mental and bodily labour of more than three years, has presented to the public notice, in the gardens of the Crystal Palace at Sydenham, restorations of no less than thirty-three extinct animals, known to us only by their fossil remains,—I am enabled to give an original sketch, from his own pencil, of his marvellous models of ancient marine Saurians, the originals of which are now at Sydenham.

Guiding his artistic hand by the descriptions of the anatomy of these marine monsters, and by the deductions therefrom relative to their habits and mode of life, as explained by Cuvier, Professor Owen, Dr. Buckland, Dr. Mantell, and others, he has again, so to speak, clothed their dry bones with skin and muscle, and placed them in the attitudes which in all probability they assumed when in life. The whole group is calculated to give an idea of scenes which took place at such an exceedingly remote period of the earth's history, that the attempt to realize the incalculable ages which have since passed away, staggers and confuses the limited conception of the human intellect.

Francis Buckland, note in new edition of William Buckland's "Bridgewater Treatise," *Geology and Mineralogy* (1858).

Text 54

The rapid exhaustion of the first edition of this work has made a second one necessary. The goodness of the public obliges me to make every effort to make these scenes, which I offer to the friends of palaeontology, as complete as possible.

Although the past decade has been rich in palaeontological observations and publications, the science itself has not changed radically. The fourteen scenes of the first edition can therefore be republished without needing any significant changes in the text. However, since our knowledge of the primordial state of the plant world has made great progress, I have thought I should add two supplementary Plates, which form as it were the prologue to this drama in which the history of creation unfolds. These new scenes will perhaps help to make the idea of the past clearer and more distinct; they will give a sense of direction to our spirit, which until now—one must recognise—has too often wandered without a guide, losing itself finally in desolating obscurity.

Thanks to the discoveries of science, we can see, right from this primitive epoch, the rule of the same laws and means that rule the present world. We are thus led to infer that these laws have been established for eternity and that they are adequate for their task!

Franz Unger, *Primitive World,* preface to second edition (1858).

Figure 66. A scene of Liassic life, drawn by Waterhouse Hawkins for the posthumous new edition of William Buckland's "Bridgewater Treatise," *Geology and Mineraology* (1858).

153

small islands have now appeared in the primeval ocean, and with them the first land plants. Although they are quite large in size, these plants are fairly primitive in form; their lack of woody tissue, Unger suggests, is the reason for the absence of coal deposits of this age. (The contrast between Unger's scene and any modern representation of the primitive land flora of the Devonian period is due to problems with dating the relevant strata: Unger's fossils were not Devonian in modern terms.) After this second new lithograph, the original opening scene, "The Transition Period" (fig. 43), takes third place and is reused without change, except that the word "Newer" is added to its title. Thereafter the sequence continues forward through deep time without further modification.[13]

The progression and diversification of life portrayed in Unger's sequence, now extended back in time by these two additional scenes, still did not strictly demand any specific causal theory. But such a sequence, and the paleontological research on which it was based, was certainly becoming persuasive evidence in favor of some kind of evolutionary explanation. Of such theories there was no shortage around mid-century, even if most of them lacked clarity of formulation. When Darwin published his *Origin of Species* (1859), it seemed at first to be just another theory to add to the many already available, none of which was generally considered satisfactory.[14] But one novel feature was its emphasis on what was soon dubbed the "struggle for existence," in contrast to the more traditional emphasis on the harmony of the natural world.

That theme was echoed in the title of a large colored lithograph that Hawkins produced the year after Darwin's celebrated book appeared. Abandoning the Secondary reptiles for a change, he drew the scene "Struggles of Life among the British Animals in Primaeval Times." It depicted the geologically recent fauna that Cuvier's and

Text 55
THE SILURIAN PERIOD

Having maintained its state of incandescent fluidity through an immeasurable period of time, the terrestrial globe cooled sufficiently to allow the formation of a solid crust at its surface. The fluid that covered this crust in the form of vapour condensed and formed a broad limitless ocean, broken by no continent or island—a bleak desert of water in which as yet no living being could exist.

The formation of vapours alternated with the deposition of precipitates. The gaseous atmosphere of the earth, and at the same time its temperature, underwent an unknown series of variations.

At last the light came to pierce this dark envelope, and its life-giving rays reached the surface of the water. The conditions for organic creation were realised, and it duly appeared. As far as our researches into the primitive state of organic creation allow us to infer, plants and animals were created together. The aqueous element was necessary for both, and the rays of the sun lit the flame of life in them.

This primitive world with its strange developments is displayed to us in the most ancient of the fossiliferous strata, which are scarcely found in Europe but which are widely distributed in North America. These plants, which are without analogues in the present creation, are predominantly aquatic. We can only link them to our seaweeds; but they differ from them in such a way that we are led to recognise in them only the ancestors of all those forms that were successively developed later, including those existing now. The same hypothesis can be applied to the animals, among which the forms then dominant were the lowest in the series, such as corals, molluscs and crustaceans etc.

Since our eye cannot penetrate into the depths of the sea, this scene [fig. 67], in order to give a visual representation of the state of nature, can only show us a limitless aqueous desert, above which is a sky that, as it were, supports the columns of vapour.

This veil, rent apart here and there, allows us to infer the presence of the divine breath on the waters.

The sea is nowhere of great depth. The tides are established and allow us to see some underwater reefs. On one of the nearer masses we notice some of the algae that we know as *Palaeophycus, Buthotrephis, Harlania* and *Sphenothallus;* among them is a multitude of corals, graptolites, orthoceratites, crinoids, brachiopods and other molluscs. At the same time trilobites resembling the crabs of the Moluccas or the *Serolis* of the Pacific take refuge in the water,

Figure 67. "The Silurian Period": the first new lithograph by Josef Kuwasseg in the second edition of Franz Unger's *Primitive World* (1858).

155

Buckland's research had begun to reconstruct forty or fifty years earlier. This was the fauna that was now attributed to a relatively cold "glacial" or "Pleistocene" period. Following the standard conception of that period, Hawkins drew a wholly prehuman scene. It was the largest two-dimensional representation of a scene from deep time that had yet been published—the lithograph was almost three feet across—and it was clearly intended to serve as a visual aid in classrooms and lecture halls.

Hawkins's "Struggles of Life" was put on sale by James Tennant (1808–81), a London mineralogist and geological lecturer, and the proprietor of a flourishing business that catered for collectors of minerals and fossils. Even the small keyed sketch (fig. 69) that served to advertise this wall-chart and its accompanying explanation (text 57) are sufficient to suggest the lively way that Hawkins showed the Pleistocene mammals in intensive interaction: eating and being eaten, chasing and fleeing, and with a sabre-toothed tiger getting its surprised comeuppance in the trunk of a mammoth.[15] This is no longer the Arcadian world of many earlier scenes from deep time, but nor is it the nightmarish or apocalyptic vision of Martin's designs. Instead it shows the "Nature red in tooth and claw" that Darwin's evolutionary theory was popularly supposed to entail.

Soon afterward, and perhaps as a result of the success of this large scene, Hawkins was commissioned by the Department of Science and Art—a governmental body set up after the Great Exhibition of 1851, to continue its educational work—to draw a whole set of such scenes. These would make his reconstructions available, not least in school classrooms, to those unable to visit them at the Crystal Palace. Tennant was in fact already offering Hawkins's Secondary reptiles for sale in another form, as small-scale three-dimensional models.[16] But Hawkins now grouped his reptiles in three scenes of Secondary life

for their eyes cannot cope with light that is not tempered by a medium denser than air.

In the distance a series of rocks emerge from the sea, with masses of seaweeds analogous to those that today form the Sargasso Sea.

Franz Unger, *Primitive World*, second edition (1858).

Text 56
THE DEVONIAN PERIOD

The continents have been permanently elevated above the waters, but they are still small and form scattered islands. Proportionately, the ocean has here and there increased in depth. Land and sea have become a workshop in which new and varied organisms are formed.

It is interesting to glance at the island vegetation of this period [fig. 68]. The earth, rich in productive means, is covered as if by magic with a carpet of vegetation that no mortal sight can contemplate, the existence and form of which is only revealed to us by some mutilated fragments. Despite their differences from our present vegetation, one can nevertheless recognise a certain match in their fundamental forms and character. There are leafy trunks, which grow in clusters on a soil that is damp or made marshy by the stagnant waters. All the leaves of these plants are small, without support, or there are even merely girdles of leaves, which generally surround the branches or trunks. These scaly or linear leaves indicate plants that are incomplete, and if several are the size of trees, their slender and poorly developed trunks suggest great fragility. Numerous aerial roots fix the stems to the soil. It is as if one were looking into a thicket of gigantic mosses. Some of these plants are well enough preserved that one can recognise their original forms, their habit and their nature. Just from the anatomical structure of the stems we know that analogous forms are lacking in our present flora, and that the most characteristic of these plants represent a combination of the

mosses and the leafy Lycopodiacea [clubmosses] and Equisetacea [horsetails]. The plants depicted in this scene should therefore be regarded as intermediates between these diverse forms. Thus the group of trees to the left consists of *Cladoxylon mirabilis*, beneath which are shrubs of *Asterophyllites coronata*, while the rest of the forest trees extending to the right foreground belong to *Schizoxylon taeniatum*, a highly enigmatic plant. The extensive rhizome-like stems have their analogue in many mosses and lycopods, and contribute essentially to give them a distinctive character.

Finally, one can see further to the left and in

Figure 68. "The Devonian Period": the second new lithograph by Josef Kuwasseg in the second edition of Franz Unger's *Primitive World* (1858).

(figs. 70–72) and added two scenes of Tertiary mammals and a final scene of those of the "Post-Tertiary epoch," or Pleistocene (figs. 73–75). Like his "Struggles of Life," Hawkins's set for the Department of Science and Art was sold by Tennant.[17]

These six designs show little of the ecological dynamics that had made Hawkins's earlier rendering of the Pleistocene so lively. Like his exhibit at the Crystal Palace, most of the animals pose in a series of static tableaux. The aquatic Liassic reptiles, and the crinoid from the same period, are even shown unrealistically out on dry land (fig. 70). Nonetheless, these designs must have helped to propagate—not least among the children in British schools—the notion that these spectacular extinct animals had not all lived at the same epoch. As Unger's series demonstrated in another format, deep time had been highly differentiated. (Hawkins later moved to the United States and started work on an abortive exhibit for Central Park in New York, which would have displayed this diversity in the same format as at the Crystal Palace, while also including the first models of the newly discovered American dinosaurs. His fine mural paintings of a sequence of scenes survives, however, at Princeton University.)[18]

Implicit in Hawkins's final wall chart (fig. 75), as in his earlier scene of Pleistocene life (fig. 69), was the assumption that no human beings had been present at the time of these extinct mammals. But this was just what had become increasingly doubtful, or at least controversial, by the time Hawkins drew those scenes. The debate on "the antiquity of man," as it was termed, was nowhere more lively, indeed vehement, than in Paris, where the posthumous authority of Cuvier continued to cast a long shadow of skepticism over any discovery claims involving human fossils. However, a suspicion that early human beings might have coexisted with Cuvier's fauna of large

front a group of almost herbaceous plants. Their fruits have a strong resemblance to the capsules of mosses with their large apophyses, and they can be taken truly as mosses of arborescent structure; this supposition is confirmed by the discovery of the structure of mosses in *Aphyllum paradoxum*.

This forest of *Cladoxylon*, with the character of a luxuriant coastal vegetation, is developed without the influence of an intense sun or of humid mists, above all in the parts that are most dense and inaccessible to light. Judging by the spongy texture of the trunks and the complete absence of wood, their decomposition, like their growth, would have been very rapid. This is why this vegetation was not favourable to any lasting accumulation of coaly matter.

Franz Unger, *Primitive World*, second edition (1858).

Text 57

STRUGGLES OF LIFE AMONG THE BRITISH
ANIMALS IN PRIMAEVAL TIMES

Elephants, Lions, Tigers, Rhinoceroses, Hippopotami, Reindeer, Elks, Bears, Hyaenas, and Wolves, were all among the native inhabitants of this land, which we call Great Britain, and were living in vast numbers in this country during a long period of time, as is satisfactorily proved by the quantity of their bones found in the superficial deposits . . .

Mr. Waterhouse Hawkins, so well known for his successful realizations of the restored forms of the extinct animals in the Crystal Palace Park at Sydenham, has endeavoured in this picture [fig. 69], called the "Struggles of Life," to represent one of those encounters which must have so frequently occurred in the evenings of those days of drought when the gigantic Mammoth, great-horned Elk, Reindeer, and various bovine animals would naturally seek the margins of lakes and river-pools. Such a country as some of our mountain-limestone districts must have abounded with fissures and caverns

continued on page 166

KEY TO A COLOURED LITHOGRAPHIC PLATE OF

WATERHOUSE HAWKINS'S RESTORATIONS OF EXTINCT ANIMALS.

(Size of Print 34 in. by 28 in.)—Published by JAMES TENNANT, 149 Strand, London, W.C.—Price 12s.

STRUGGLES OF LIFE AMONG THE BRITISH ANIMALS IN PRIMÆVAL TIMES.

Figure 69. "Struggles of Life among the British Animals in Primaeval Times": a sketch to act as a key to a lithographed wall poster (not reproduced here) by Waterhouse Hawkins (1860). The animals are identified as follows:
1. Mammoth (*Elephas primigenius*);
2. Scimitar-toothed Lion (*Machairodus latidens*); 3. Cave Tiger (*Felis spelaea*); 4. Great Hippopotamus (*Hippopotamus major*); 5. Two-horned Rhinoceros (*Rhinoceros tichorhinus*); 6. Irish Elk (*Megaceros Hibernicus*); 7. Musk Ox (*Ovibos muschatus*); 8. Wolf (*Canis lupus*); 9. Cave Bear (*Ursus spelaeus*); 10. Fossil Aurochs (*Bison priscus*); 11. Fossil Ox (*Bos primigenius*); 12. Long-fronted Ox (*Bos longifrons*); 13. Cave Hyaena (*Hyaena spelaea*); 14. Reindeer (*Cervus tarandus*); 15. Red Deer (*Cervus [Strongyloceros] spelaeus*); 16. Wild Hog (*Sus scrofa*); 17. Fossil Horse (*Equus plicidens*); 18. Fossil Ass (*Asinus fossilis*); 19. Young Elephant (*Elephas primigenius*); 20. Common Otter (*Lutra vulgaris*).

159

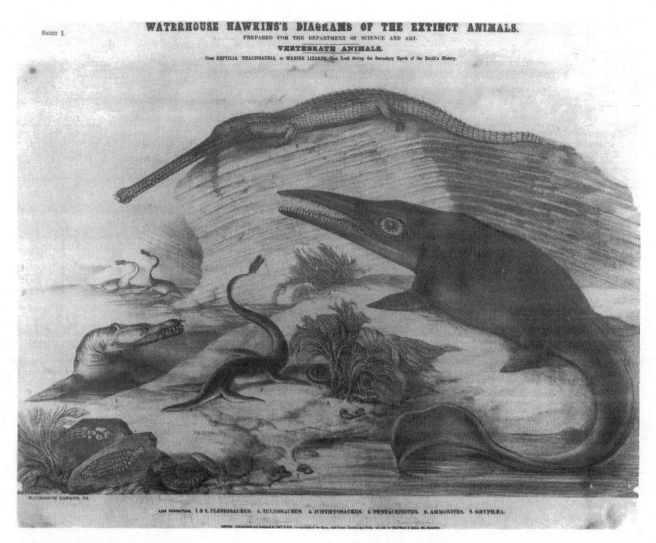

Figure 70. "Enaliosauria, or Marine Lizards that lived during the Secondary Epoch of the Earth's History": the first of Waterhouse Hawkins's set of six wall posters for the Department of Science and Art (c. 1862).

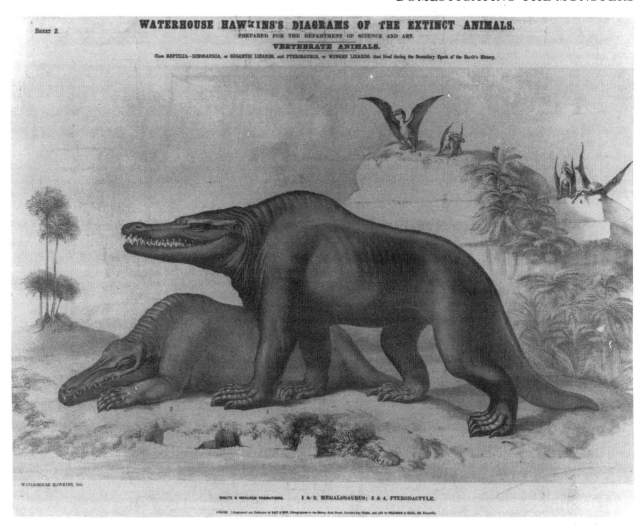

Figure 71. "Dinosauria, or Gigantic Lizards, and Pterosauria, or Winged Lizards, that lived during the Secondary Epoch of the Earth's History": the second of Waterhouse Hawkins's set of wall posters.

Figure 72. "Dinosauria, or Gigantic Lizards, that lived during the Secondary Epoch of the Earth's History": the third of Waterhouse Hawkins's set of wall posters.

Figure 73. "Pachydermata, that lived during the Tertiary Epoch of the Earth's History": the fourth of Waterhouse Hawkins's set of wall posters.

Figure 74. "Edentata, that lived during the Tertiary Epoch of the Earth's History": the fifth of Waterhouse Hawkins's set of wall posters.

Figure 75. "Pachydermata and
Carnivora, that lived during the
Post-Tertiary Epoch of the Earth's
History": the last of Waterhouse
Hawkins's set of wall posters.

extinct mammals had persisted, despite the disapproval of the scientific establishment. Indeed, it had grown, particularly as a result of the long-continued work of Jacques Boucher de Perthes (1788–1868) on the "diluvial" gravels of the river Somme near his home at Abbeville in northern France. Boucher, a customs official and amateur prehistorian, had long claimed to have found chipped flint implements of apparently human workmanship in direct association with the familiar bones and teeth of mammoths and other extinct mammals. But he had set his claims in the context of theories of human prehistory so bizarre—they bore some resemblance to those of Isabella Duncan—that they were readily dismissed. Only in the years around 1860 had the Parisian establishment begun to take Boucher's claims more seriously.[19]

Such claims were first given pictorial expression in a popular book entitled *Paris before Men* (*Paris avant les hommes*, 1861). This was a posthumous publication by the French botanist and geologist Pierre Boitard (1789–1859). In order to introduce his readers to the idea of deep time, Boitard uses the literary device of an explicitly magical or fairy-tale character. He conjures up Asmodée, the lame demon (*le diable boiteux*—in French a nice pun on his own name), whom he borrows from Le Sage's classic novel of that name (1707), to conduct him on his adventure. One of the first illustrations (not reproduced here) shows the two, sitting comfortably on a large meteorite as if on a Paris omnibus, traveling through deep time as if through deep space.[20]

The very first of Boitard's true scenes from deep time therefore depicts these two characters—the human and the magical—as actors within the scene itself, just as Buckland had been an actor within his den of extinct hyenas (fig. 17). The magic has whisked the demon in period costume and the elegantly dressed Parisian back in deep time into the world of the plesiosaur (fig. 76;

that formed dens and dwelling-places for numberless troops of Wolves, Hyaenas, and the larger carnivora, all dependent for their food upon the success of their attacks on the gigantic herbivora seeking the water-places, and straying into the neighbourhood of the dens of these their natural enemies.

We may therefore readily imagine troops of ravenous Wolves, driving whole herds of Elks and Deer, with individuals of the larger cattle, across the path even of the giant Mammoth, who thus becoming entangled in the *melée*, panic-struck, and attacked on all sides, would fall an easy prey to the fangs of the Cave Lion or Scimitar-toothed Tiger.

The slow but powerful Bears, with the fierce and strong Hyaenas, were there, ready to share the spoil of the more active hunters, and drag into their dens such portions of the carcases as their strength and instinct enabled them to remove. The well-gnawed bones of the victims and the polished corners of the caverns are among the evidences of the cave-habits of the old Hyaenas and their contemporaries. Thus aided by the general struggle, the larger and more powerful Felidae (the Lion and Tiger) might well succeed in overpowering the gigantic Mammoth, Rhinoceros, and Hippopotamus,—though these thick-skinned animals might be supposed to be invulnerable to the attacks of these formidable carnivora, did we not find that some of the latter were especially armed to penetrate the skins of these giants, as shown by the scimitar-teeth of Machairodus, which is represented as writhing in the strangling coil of the Mammoth's trunk: but a more effective attack is made by other individuals of the same species on the great Hippopotamus.

The unarmed Horse and timid Deer, mired in the swamp or crushed in encountering herds, would fall an easy prey, and, together with the weak and young of every kind, would often choke the ravine and the morass with heaped-up carcases. Hence we find so frequently cave-earth and breccia thickly crowded with impacted bones, exhibiting in the fossil state the broken remnants of once living multitudes.

James Tennant, pamphlet advertising Waterhouse Hawkins's large lithograph (1860). Although Tennant is identified as the author of the pamphlet, the text may have been written for him by Hawkins.

Text 58

I saw again a fern of the preceding period, and I wanted to get nearer this tree, the roots of which were in the water; I was already putting out my hand to pluck a leaf, which I intended for the herbarium of the Museum of Natural History in Paris, when an acute menacing whistle could be heard nearby. I recoiled in terror on seeing the scaly head of a horrible reptile looking at me with flashing eyes. Its open mouth with sharp teeth menaced me with a forked sting; its neck was of a prodigious length, like a cable, or rather like a huge snake; its massive body, covered with large yellowish scales, was rather like that of an enormous fish; but it had four short legs, of which the digits were covered with a thick membrane, which gave them some resemblance to those of a sea turtle; the short stout tail of a crocodile served it as a rudder.

"It's a plesiosaur," said the genie.

—It's a strange monster, the form of which is so fantastic that, if I hadn't seen it with my own two eyes, it would seem the product of the delirious imagination of a poet, rather than of the hand of nature.

Pierre Boitard, *Paris before Men* (1861).

text 58). The nightmarish horror of the encounter, heightened by such gratuitous details as the reptile's forked tongue and yellowish scaly coat, establishes the required tone of monstrosity in the ancient world. With the principle of magical time-travel thus established, Boitard

Figure 76. "Plesiosaur": the first scene from deep time in Pierre Boitard's *Paris before Men* (1861). The narrator has been transported by the magic of his companion, the "lame demon" (*diable boiteux*), back into the time of the Liassic reptile. This and the other illustrations in the book were drawn by Boitard himself and engraved by Moreau.

167

does not bother to depict himself in any of the later scenes.

The frontispiece, however, is significant: it is a highly unflattering and monkeylike representation of "Fossil Man" (fig. 77), wielding a stone axe against unseen enemies and defending his equally simian mate and offspring at the mouth of their cave. This design showed Boitard's readers at once where he stood in the controversy about human origins, and the corresponding nar-

Text 59

"Are you afraid?" the demon asked me.

—I believe we are going to encounter animals even more formidable than those we met on our way here.

But the genie threw me such a forcefully ironic glance that I was ashamed of my weakness, and I entered the cave with a determined step. . . . Little by little my pupils dilated, and I was able to see, vaguely at first, the objects that surrounded us: a hyaena, with its skull split as if it had been struck on the head with an axe, was stretched out at our feet, and several scraps of bear's flesh, half eaten, were strewn here and there on the ground, exuding a highly unpleasant smell. . . . But what astounded me most was a kind of clay pot, not fired but sun-baked, very crudely made, and half full of the still warm blood of the hyaena. The genie pointed out that on the edge of the pot were the bloody marks of lips that had drunk the disgusting liquid it contained. By the side of the pot I saw a fragment of flint, trimmed roughly into the form of a tapered axe, mounted at the end of a stick, and bound firmly with strips of bear's skin. This instrument was closely similar to the tomahawk of the Canadian savages. . . .

The genie put his finger to his mouth, signalling me to keep silent and to move forward with care; which I did. Then he gently lifted the bear skin and revealed to my eyes the most singular and horrible animals I had seen until now. There were three of them, two large, and a small one that I recognized as the young of this horrible species. . . . Its body had rather the form of an orang-utang, but without being either nimble or graceful, because it was stout, squat and thickly muscular . . .

Pierre Boitard, *Paris before Men* (1861).

Figure 77. "Fossil Man": the frontispiece of Pierre Boitard's *Paris before Men* (1861).

Figure 78. "Anthropic Period; Last Palaeontological Age; Appearance of Man": the final scene in Pierre Boitard's *Paris before Men* (1861).

rative (text 59) accentuates the bestiality of his readers' forebears. At the end of the book, his final scene depicts what he terms the "Anthropic Period," setting those hardly human beings unambiguously in a landscape of extinct mammals (fig. 78).

Boitard's poor illustrations—they were engraved from his own drawings—and perhaps also his crudely polemical position on human ancestry, were probably enough to convince another Parisian popularizer of science that he could do much better. Certainly another factor in that author's decision to write a similar book was his discovery of Unger's sequence, in the enlarged edition described earlier. He was also evidently aware of Hawkins's work, either from a visit to the Crystal Palace (or at least from

169

French newspaper reports and illustrations of the exhibit there), or from Hawkins's set of large wall charts. This confluence of pictorial resources led to the publication of a sequence of scenes from deep time that was so influential for the future of the genre that, like Unger's original sequence, it deserves a chapter to itself.

Since Kuwasseg's sequence of scenes for Unger was not initially well known, many popular books on geology around mid-century continued to show only a single scene, usually centered on the Liassic (Jurassic) reptiles, which purported to characterize the whole of "the ancient world." Few designs attempted to convey a sense of temporal change or progress. Duncan's illustration of her Pre-Adamite theory—one of the exceptions—displayed a total discontinuity between the present world and its predecessor. As before, the genre of scenes from deep time showed itself to be highly flexible and capable of being adapted to a wide variety of theoretical messages.

The anti-Lamarckian message that Owen wanted to inject into Hawkins's exhibit at the Crystal Palace was probably decoded by very few among the crowds that went to gaze at the huge reconstructed reptiles. Even the message of temporal sequence, which was built into the spatial arrangement of the models, was probably overlooked. What remained was simply an impression of a single "antediluvian" world, inhabited by creatures that were perceived as "monsters" fit for a nightmare. But the Crystal Palace exhibit was much more than just a blow-up of the stereotyped single scene that had become a conventional feature in many popular books on the "wonders" of geology. The impact of three-dimensional reconstructions, modeled at full size, can hardly be overestimated. These were creatures that were believable to the lay eye, almost as much as the exotic living animals on display at the zoo in central London. Hawkins's exhibit brought the scientists' vision of the reality of the deep past to the imagination of a mass public as never before, and it did so with the element of showbiz that has been inseparable from dinosaur displays ever since. Even the limitation entailed by the immobility of the exhibit was alleviated, and the virtual audience for the show still further enlarged, by the marketing of small-scale models and by the large wall charts that decorated classrooms and lecture halls far beyond London.

What made "the ancient world" of the extinct reptiles so monstrous and alien was above all that it was totally lacking in the human presence. Indeed, it was generally assumed that even the much more recent world of the so-called diluvial animals—usually attributed by now to some kind of glacial period—had also been wholly pre-human. But that assumption had begun to be seriously questioned, and the possibility that human beings too had belonged in "the ancient world" could no longer be dismissed. Boitard's otherwise undistinguished popular book was the first to make visible—in the vivid form of a scene from deep time—what such an embedding of human nature into the history of nature might signify.

Boitard's book was also important for making explicit what any scene from deep time entails, namely, some principle of time travel, by which modern human beings can be transported, at least in imagination, back into a scene that no human beings can really have witnessed. But it is significant that, just as earlier scenes had suggested the medium of the fairy tale or the dream, Boitard made his time travel explicitly magical, drawing on traditional resources from his own literary culture. He had lived through a period of unparalleled technological progress, symbolized above all by the previously inconceivable speed of travel made possible by the train. But

171

Boitard could not imagine—or thought his readers would be unable to imagine—the idea of any kind of time *machine* as even a fictional means of traveling back into prehuman history. The reconstruction of the deep past, although hailed as one of the "wonders" of *science*, still had to depend on far more traditional resources to be made comprehensible and persuasive to the general public.

6

The Genre Established

Guillaume Louis Figuier (1819–94) trained and qualified as a physician at Montpellier. He then pursued chemical research in Paris, and in 1853 he was appointed a professor at the school of pharmacy there. By then he had begun to turn his talents to the popularization of science. His *Great Discoveries of Modern Science* (*Exposition et histoire des principales découvertes scientifiques modernes*, 1851) was so well received that François Arago, by then a distinguished elder statesman of French science, advised him to concentrate on such work. This he did, producing, for example, four volumes of *Marvels of Modern Times* (*Histoire des merveilleux dans les temps modernes*, 1860).[1] As already suggested, it may have been Boitard's posthumous book on the history of life, published the following year, that made Figuier decide to write his own book on the same theme. It must have been around that time that he started work on *The Earth before the Deluge* (*La terre avant le Déluge*), which appeared in 1863 and was an immediate success.[2]

Like all Figuier's popular books, this one was profusely illustrated. Most of the illustrations were borrowed from a respectable academic textbook: the two-volume *Ele-*

mentary Course on Palaeontology and Stratigraphical Geology (Cours élémentaire de paléontologie et de géologie stratigraphique, 1849–52) by Alcide d'Orbigny (1802–57), who until his death had been professor of paleontology at Cuvier's institution, the Museum of Natural History in Paris. Figuier took from d'Orbigny's work some three hundred engravings of fossils, including a few reconstructed skeletons with inferred body outlines, of the kind that Cuvier had pioneered half a century earlier (fig. 13). But there were no scenes from deep time to be had from d'Orbigny's book.[3]

Figuier evidently decided that such scenes, if artistically and scientifically superior to Boitard's, would greatly improve the attractiveness of his book for the general public and, not least, for the young readers at whom it was specifically directed.[4] In particular, Figuier was aware of the great series of scenes that Kuwasseg had designed for Unger, and explicitly took them as his model. He therefore enlisted the help of a young Parisian landscape painter and illustrator, Edouard Riou (1833–1900), for this project. Riou drew more than two dozen full-page illustrations for Figuier, most of them scenes from deep time. They were important enough in the marketing of the book for his artistic contribution to be mentioned prominently on the title page. (At around the same time, Riou also began illustrating the work of the other great French popularizer of science of this period, Jules Verne.)[5]

Unger had extended his sequence of scenes backward into still deeper time when he added two new plates to his second edition (figs. 67, 68). Figuier continued this trend by beginning his series not just with the earliest known organisms, but with a period before the origin of life itself (fig. 79; text 60). Figuier's first scene is wholly conjectural, yet based on the standard geological theory of the time: the earth is assumed to have begun its long history as an

Text 60

The first water which fell in the liquid state upon the gradually cooling surface of the earth would be rapidly reduced to steam by the elevation of its temperature. Thus rendered much lighter than the surrounding atmosphere, these vapours would rise to the utmost limits of the upper atmospheric zone: thus circumstanced, they would radiate towards the glacial regions of space, and, again condensing, they would again descend to the earth in a liquid state, to reascend as vapour and fall again in a state of condensation. But these alternate changes in the physical condition of water could only be maintained by a very considerable temperature on the surface of the globe, which these alternations of heat and cold were rapidly diminishing: the excess of heat was being dissipated in the regions of celestial space.

This phenomenon extending itself by degrees to the whole mass of watery vapour existing in the atmosphere, the waters in increasing quantities covered the earth; and as the conversion of all liquids into vapour is provocative of a notable disengagement of electricity, a vast quantity of electric fluid necessarily resulted from the conversion of such masses of water into vapour. Bursts of thunder, and bright gleams of lightning, were the necessary accompaniments of this extraordinary struggle of the elements [fig. 79].

Louis Figuier, *Earth before the Deluge* (1863).

Figure 79. "Condensation and
Rainfall on the Primitive Globe": the
first of Edouard Riou's engraved
scenes in Louis Figuier's *Earth before
the Deluge* (1863).

Figure 80. "Ideal View of the Earth during the Silurian Period," from Louis Figuier's *Earth before the Deluge* (1863).

Text 61

Figure [80] represents an ideal view of the earth during the Silurian period. Immense shallow seas expose, here and there, underwater reefs covered with algae and frequented by various molluscs and articulates [trilobites]. A pale sun, which pierces with difficulty the heavy atmosphere of the primitive world, shines on the first living beings to leave the hands of the Creator, often with rudimentary organization, but in other cases advanced enough to indicate progress towards more perfect beings.

Louis Figuier, *Earth before the Deluge* (1863).

Text 62

Vast seas, covered with a few islets, form the ideal of
the Devonian period. Upon the rocks of these islets
the mollusks and articulata of the period exhibit
themselves [fig. 81]. Stranded on the shore we see a
cuirassed fish, of strange form. A restored group of
shrubs, *Asterophyllites coronata,* covers one of the islets, mixed with plants nearly herbaceous, resembling mosses, though the true mosses did not appear
till much later. *Encrinites* and *lituites* occupy the
rocks in the foreground on the left hand.

The vegetation is still humble in its development,
for forest-trees are altogether missing. The asterophyllites rise singly to a considerable height, with tall

Figure 81. "Ideal View of the
Earth during the Devonian Period,"
from Louis Figuier's *Earth before
the Deluge* (1863).

177

incandescent ball in space. Riou depicts torrential rains falling onto a primeval ocean, as the still hot globe begins to cool enough to condense the water vapor out of its original atmosphere.

With the second scene, life makes its first appearance (fig. 80; text 61). Riou's representation of the Silurian period is obviously inspired by the first of Unger's new scenes (fig. 67). The sun hardly pierces the still dense atmosphere; a few islands have emerged, but there is no life on dry land; the marine organisms, trilobites conspicuous among them, are shown in the conventional manner, thrown up on the shore.

Figuier's third scene, however, shows he is not following Unger slavishly. Riou's picture of the Devonian period (fig. 81; text 62), unlike Unger's second new scene (fig. 68), shows animals as well as plants. The animals are indeed laid out neatly on the rocky shore, looking more like specimens in a museum cabinet than storm-tossed debris; but they do at least acknowledge the rich fossil fauna of the period. The armored fish in the foreground represents the first appearance of vertebrates (the Old Red Sandstone, from which such fossils came, was at this time regarded as a marine deposit). Primitive plants similar to those depicted in Unger's scene are here demoted to a spur of land in the middle ground but are important for indicating that life has now spread to dry land.

The scene illustrating the period of the next major formation, the Carboniferous Limestone, exists in two versions, with an important pictorial difference. Riou originally drew a shoreline scene in which the rich marine fauna was shown in the usual way, thrown up on the beach, and quite overshadowed by trees and other plants like those of the slightly later Coal formation (fig. 82; text 63). But by the fourth edition, only two years later, Figuier had replaced this with a scene that was explicitly modeled on a view of an aquarium (fig. 83; text 64). De la

and slender stem. The light, still pale, seen through the semi-opaque atmosphere, only permits of a vegetation essentially cellular, sluggish, and vascular. Cryptogames, of which the mushrooms convey some idea, would be the chief vegetation; but in consequence of the softness of their tissues, their want of consistence and of woody fibre, no vestiges of them have come down to us.

Louis Figuier, *Earth before the Deluge* (1863).

Text 63

[Fig. 82] shows the various natural elements that belong to the sub-period of the Carboniferous limestone, assembled in an idealised manner. One sees a series of islands raised above a calm sea and covered with the vegetation proper to this period.

The stout trunk that closes the landscape on the left is that of a *Sigillaria*. On the ground, one notices the fruit of this large plant, with the form of long imbricate ears. Near the trunk of the *Sigillaria* is a *Lepidodendron*, with its foliage; some herbaceous ferns are seen at the base of the *Lepidodendron*. On the islet to the right, between two *Calamites*, rise the stems of the *Lomatophloios*, the branches of which resemble feather dusters.

On the beach, at the moment when the tide is turning, the molluscs, crustaceans and zoophytes proper to this period are shown. In the background one sees an elevation of the ground, accompanied by an emission of gas and vapours; a geological phenomenon that was frequent at this remote period in the evolution of the earth.

Louis Figuier, *Earth before the Deluge* (1863).

178

Figure 82. "Ideal View of the Earth during the Period of the Carboniferous Limestone," from Louis Figuier's *Earth before the Deluge* (1863).

Beche's innovation of more than thirty years earlier (fig. 19) was now at long last adopted in at least one scene within a major series: with the craze for seawater aquaria in full swing, such a view was now familiar to the readers of popular books such as Figuier's. The aquariumlike design brings the animals together in close-up, but also in an implausible concentration; the plant life of the period is now relegated to the background.

The next scene, however, depicts the Coal period, and the plants occupy almost the whole stage. Riou first drew a scene (not reproduced here) that, although entitled "Swamp of the Coal Period," portrayed a landscape more like a savannah, with the characteristic Coal Measures trees scattered across ground with grass-sized plants. But this was replaced soon afterward by a more appropriate scene of a dense and gloomy forested swamp (fig. 84; text 65). This is clearly based on Kuwasseg's similar scene (fig. 44), although the small steel engraving lacks the detail and atmospheric subtlety of the large lithograph. Some fish swim in clear water in the foreground, recalling those in Goldfuss's scene (fig. 40). But more recent discoveries allow Riou to show in addition the earliest reptile, *Archegosaurus,* also swimming (since only its head was yet known, the water at this point is conveniently *not* transparent!).

Riou's scene of the Permian period (fig. 85; text 66), is even more clearly inspired by Kuwasseg's earlier lithograph for Unger (fig. 46). The volcanic activity known from this period in some parts of Europe is again made responsible for its characteristic landscape: the eruptions in the background create a locally stormy scene. In the foreground are trees and other plants similar in general effect to Unger's; but Riou adds a sample of marine life, though he reverts to the traditional manner of displaying them.

For Figuier's next scene, representing the earlier part

Text 64

[Fig. 83] represents—thanks to the artifice of a sort of ideal aquarium—some of the more prominent species which inhabited the seas during the carboniferous age. On the right is a tribe of corals, with reflections of dazzling white: the species represented are, nearest the edge, the *Lasmocyathus,* the *Chaetitus,* and the *Ptlypora.* The mollusk which occupies the extremity of the elongated, conic and sabre-like tube—an *Aploceras,* seems to prepare the way for the ammonite; for if this elongated shell was rounded and turned round its centre it would approximate to the ammonite and nautilus. In the centre of the first plane we have *Bellerophon huilcus,* the *Nautilus Koninckii,* and a *Productus,* with the numerous spines which surround the shell.

On the left are other corals: the *Chonetas* at the surface, extended and furnished with small spines, and the *Cyathophyllum,* with straight cylindrical stems, some Encrinites, *Cyathocrinus* and *Platycrinus,* rolled round the trunk of a tree, or with their flexible stem floating in the water. Some fishes, *Amblypterus,* move about in the middle of these creatures, the most part of whom are immoveably attached, like plants, to the rock on which they are rooted.

In addition, this engraving shows us a series of islets, rising above the tranquil sea. One of these is occupied by a forest, in which a distant view is presented of the general form of the grand vegetation of the period.

Louis Figuier, *Earth before the Deluge,* revised version in fourth edition (1865).

Text 65

[Fig. 84] is an attempt to represent a marsh and a forest of the Coal period. One sees a short and thick vegetation, a sort of grass, composed of herbaceous ferns and mares-tail. Several trees of forest height raise their heads above this lacustrine vegetation. Here are the indications of the species represented:—

On the left are seen the naked trunk of a *Lepido-*

Figure 83. "Marine Animals of the Carboniferous Limestone Period," revised version, in fourth edition of Louis Figuier's *Earth before the Deluge* (1865).

dendron and a *Sigillaria,* an arborescent fern rising between the two trunks. At the foot of these great trees an herbaceous fern and a *Stigmaria* appear, whose long ramification of roots, provided with reproductive spores, extend to the water. On the right the naked trunk of another *Sigillaria,* a tree whose foliage is altogether unknown, a *Sphenophyllum* and a *conifer.* It is difficult to describe with precision the species of this family, whose imprints are, nevertheless, very abundant in the coal formation.

In front of this group we see two trunks broken and overthrown. These are a *Lepidodendron* and *Sigillaria,* mingling with a heap of vegetable débris in course of decomposition, from which a rich humus will be formed, upon which a new generation of

Figure 84. "View of a Forest and Swamp during the Coal Period," revised version, in fourth edition of Louis Figuier's *Earth before the Deluge* (1865).

plants will soon develop themselves. Some herbaceous ferns and buds of *Calamite* rise out of the marshy water.

A few fishes, belonging to the period, swim on the surface of the water, and the aquatic reptile *Archegosaurus* shows its long and pointed head—the only part of the animal which is known. A *Stigmaria* extends its roots into the water, and the pretty Asterophyllites rise above it in the first plane, with finely-cut stems.

A forest, composed of *Lepidodendrons* and *Calamites,* forms the background to the picture.

Louis Figuier, *Earth before the Deluge,* revised version in fourth edition (1865).

Text 66

In the background of [fig. 85], which represents an ideal view of the earth during the Permian period, one sees a mass of steam and vapour rising in columns from the water, resulting from the still-smoking and scarcely-consolidated matter. In the foreground, on the right, rise groups of tree-ferns, Lepidodendrons, and Walchias, of the preceding period. At the edge of the sea, left exposed by the retiring tide, are Mollusks and Zoophytes of the period—*Productus, Spirifers* and *Encrinites;* pretty little plants—the *Asterophyllites* of the carboniferous age are growing at the water's edge, not far from the shore. Having attained a certain height in the cooler

Figure 85. "Ideal View of the Earth during the Permian Period," from Louis Figuier's *Earth before the Deluge* (1863).

183

of the Trias period (fig. 86; text 67), Riou ingeniously combines two scenes, one terrestrial and one marine, from Unger's sequence (figs. 47, 48). On the left is the storm-tossed sea of the Muschelkalk, with that limestone's characteristic stalked crinoids and other invertebrates thrown up on the rocks as usual. On the right is the edge of the land, with trees reconstructed from the plant fossils from the slightly older Bunter Sandstone. The two halves of the scene are dramatically united by the confrontation between two reptiles, one borrowed from each of Unger's scenes: the terrestrial labyrinthodon, leaving its footprints on the sand, and the marine nothosaurus, virtually copied from Kuwasseg's lifelike pose (though reversed in mirror image by the engraving process).

A separate and larger picture of the labyrinthodon and its footprints (not reproduced here) is the first example of a kind of illustration that becomes increasingly common later in the book. Figuier repeatedly illustrates the inferential process that underlies Riou's full-page scenes by showing a complete chain of visual evidence. This ranges from plain specimens of fossil leaves, shells, and bones, through reconstructed skeletons with or without body outlines; these are then shown as complete individual animals and plants, portrayed as if alive. Finally, several such individual organisms are shown assembled and related to one another in a full landscape scene.

The labyrinthodon reappears in the following scene, which represents the later part of the Trias period, during which the Keuper Sandstone and its extensive salt deposits were formed (fig. 87; text 68). The animal with its famous footprints, and the tree-sized plants, are clearly derived from Kuwasseg's corresponding scene for Unger (fig. 49). This terrestrial landscape is again compressed into one half of the scene. But this time no aquatic life is shown in the other half, which is used instead to illustrate

atmosphere, the columns of steam would become condensed and finally fall in torrents of rain. The evaporation of water in such vast masses being necessarily accompanied by an enormous disengagement of electric fluid, this gloomy picture of the primitive world is lit up by brilliant flashes of lightning, accompanied by the reverberating noise of thunder.

Louis Figuier, *Earth before the Deluge* (1863).

Text 67

The accompanying engraving [fig. 86] gives an idealized picture of the plants and animals of the [Conchylian (Trias)] period. The reader must imagine himself transported to the shores of the conchylian sea at a moment when its waves are agitated by a violent but passing storm. The reflux of the tide exposes some of the aquatic animals of the period. Some fine Encrinites are seen, with their long, flexible stems, and a few Mytulus and Terebratulas. The reptile which occupies the rocks, and prepares to throw itself on its prey, is the *Nothosaurus*. Not far from it are other reptiles, its congeners, but of a smaller species. Upon the down on the shore is a fine group of the trees of the period, that is, of *Haidingeras*, with large trunks, with drooping branches and inclined foliage, of which the cedars of our age give some idea. The elegant *Voltzias* are seen in the second plane of this curtain of verdure. The reptiles which lived in these primitive forests, and which would give to it so strange a character, are represented by the *Labyrinthodon*, which descends towards the sea on the right, leaving upon the sandy shore those curious traces which have been so strangely preserved to our days, as if they were intended to answer the interrogations of science by their wonderfully-preserved vestiges.

Louis Figuier, *Earth before the Deluge* (1863).

Figure 86. "Ideal View of the Earth
during the Conchylian Period
(Trias)," from Louis Figuier's *Earth
before the Deluge* (1863).

the origin of the salt deposits. In the foreground, this scene is therefore combined with the format of a geological section, to show how the beds of salt accumulated by evaporation in enclosed lagoons.

By contrast, the next full-page engraving is so exclusively biological that it is really intermediate in form between a true landscape scene and a simple reconstruction of an animal or plant body, of the kind just described. However, since this scene does depict two animals interacting, it deserves inclusion in this sequence (fig. 88; text 69). Riou's drawing is certainly vigorous, though an ichthyosaur confronting a plesiosaur was by now a visual cliché. An ichthyosaur spouting like a whale had figured modestly in De la Beche's *Duria antiquior* (fig. 19), but Riou magnifies the spouting into sensational proportions. The accompanying text, after giving detailed reconstructions along well-worn lines, concludes with a firm rejection of any attribution of "monstrosity" to such alien creatures. The argument combines almost evolutionary language about organic progression with a reaffirmation of traditional natural theology, in a manner that was characteristic of the time.

Having made the marine reptiles of the Lias period the stars of a scene of their own, Figuier is able to devote Riou's second Lias engraving to a purely terrestrial and mainly botanical scene (fig. 89; text 70). However, a pterodactyle is highlighted as it swoops on a giant dragonfly; growing doubts about the reptile's ability to fly like a bird are reflected in Figuier's explicit comment on its gliding habit.

Riou's next scene, originally designed to illustrate the period of the Oolite limestones, or the later part of the Jurassic period (fig. 90; text 71), is very closely derived from Kuwasseg's corresponding scene (fig. 50), allowing as usual for the mirror-image reversal inherent in the engraving process. The plant and animal life is depicted

Text 68

[Fig. 87] is both a pictorial view of the earth during the Saliferous [Triassic] period, and a diagram to explain the origin of the rock salt in the Secondary formations. The theoretical section of the rocks in the foreground displays the beds of salt formed by the geological mechanism that is about to be analysed. These beds are inclined obliquely, as a result of movements of the crust after their deposition.

We have nothing particular to say about the animals that belong to the Saliferous period. One finds no debris of fossil animals mixed with the beds of rock salt and clay that comprise the Saliferous formation. The animals that peopled the shores of the seas during the Saliferous period were the same as those of the Conchylian sub-period. Since the animals that peopled these seas—we repeat—are not found in the salt and mineral beds belonging to this epoch, we have shown only the *Labyrinthodon*, the remains of which are found at the base of this formation.

Louis Figuier, *Earth before the Deluge* (1863).

Text 69

In [fig. 88] we bring together these two great marine reptiles of the Lias, the Ichthyosaur and the Plesiosaur. Cuvier says of the Plesiosaurus, "that it presents the most monstrous assemblage of characteristics that has been met with among the races of the ancient world." It is not necessary to take this expression literally; there are no monsters in nature; the laws of organization are never positively infringed; and it is more accordant with the general perfection of creation to see in an organization so special, in a structure which differs so notably from that of the animals of our days, the simple augmentation of a type, and sometimes also the beginning and successive perfecting of these beings. We shall see, in examining the curious series of animals of the ancient world, that the organization and physiological functions go on improving unceasingly, and each

Figure 87. "Ideal View of the
Earth during the Saliferan Sub-
Period (Triassic Period)," from
Louis Figuier's *Earth before
the Deluge* (1863).

of the extinct genera which preceded the appearance of man, present for each organ, modifications which always tend towards greater perfection. The fins of the fishes of Devonian seas become the paddles of the Ichthyosaurii and of the Plesiosaurii; these, in their turn, become the membranous foot of the Pterodactyle, and, finally, the wing of the bird. Afterwards come the articulated fore-foot of the terrestrial mammalia, which, after attaining remarkable perfection in the hand of the ape, becomes, finally, the arm and hand of man; an instrument of wonderful delicacy and power, belonging to an enlightened being gifted with the divine attribute of reason! Let us, then, dismiss this idea of monstrosity, which can only mislead us, and only consider the antediluvian beings as digressions. Let us look on them, not with disgust; let us learn, on the contrary, to read in the plan traced for their organization, the work of the Creator of all things, as well as the plan of creation.

Louis Figuier, *Earth before the Deluge* (1863).

Figure 88. "The Ichthyosaur and the Plesiosaur (period of the Lias)," from Louis Figuier's *Earth before the Deluge* (1863).

Text 70

[Fig. 89] represents a terrestrial landscape of the Liassic period; the trees and shrubs characteristic of the age are the elegant Pterophyllum, which appears in the extreme left of the picture, and the Zamites, which are recognizable by their thick and low trunk and fan-like tuft of foliage. The great horses-tail, or Equiseteae of this epoch mingle with the great tree-ferns and the cyprus, a conifer congenerous to those of our age. Among animals we see the Pterodactylus specially represented. One of these reptiles is seen in a state of repose, resting on its hind feet. The other is represented, not flying after the manner of a bird, but throwing itself from a rock in order to seize upon a winged insect, the dragon-fly (*Libellulae*).

Louis Figuier, *Earth before the Deluge* (1863).

Figure 89. "Landscape of the Lias Epoch," from Louis Figuier's *Earth before the Deluge* (1863).

189

in closely similar manner, down to such features as the beached ichthyosaur skeleton and the distant swanlike plesiosaur. But as usual, Riou gives the animals greater emphasis than Kuwasseg, adding, for example, a crocodile, flying pterodactyles, and a floating ammonite (though the latter are almost vanishingly small on the engraving). A more important addition is the opossumlike marsupial climbing the pandanus root, with its young out of the pouch and clinging to mother's back. This records the first known fossil mammals; their importance in the history of life is reflected in Riou's explicit use of artistic license to show the animal at greatly magnified size.

After the first edition of the book appeared, Figuier enlarged its coverage of the Oolite period by getting Riou to add two further scenes, reflecting recent research. The first represents the Middle Oolite and depicts two striking reptiles, the *Teleosaurus* and the *Hylaeosaurus* (fig. 91; text 72). One of the teleosaurs is eating a squid, while the other, overturned and dead, allows Riou to show its belly. The hylaeosaur is explicitly borrowed from Hawkins's model at the Crystal Palace.

The second of these additional scenes, based on fossils from the Upper Oolite, is devoted to the plant life of the period (fig. 92; text 73). It also depicts reconstructions of the newly discovered *Archaeopteryx,* the earliest known bird, and of the bizarre *Ramphorhynchus.*

The following scene is once again primarily of two animals, this time from the Wealden or Lower Cretaceous period (fig. 93; text 74). The poses of the warring reptiles recall those of the fearsome monsters that Martin drew for Mantell a quarter century earlier (fig. 35); but the mood of the picture is not Martinesque at all, and the botanical background is loosely based on the scene in Unger's book (fig. 51). Unlike that picture, however, the iguanodon is here attacking (and being attacked by) a

Text 71

In [fig. 90] we represent an ideal view of the earth during the period of the lower oolite. On the shore are types of the vegetation of the period. The *Zamites,* with its large trunk covered with fan-like leaves, resembled in form and bearing the Zamias of tropical regions; a *Pterophyllum,* with its stem covered from base to summit with its finely-cut feathery leaves; conifers closely resembling our cypress, and an arborescent fern. What distinguishes this sub-period from that of the lias is a group of magnificent trees, *Pandanus,* remarkable for their aerial roots, their long leaves, and globular fruit.

Upon one of the trees of this group the artist has placed the *Phascolotherium,* not very unlike to our opossum. It was the first of the mammalia which gave animation to the ancient world. The artist has here enlarged the dimensions of the animal in order to seize the form; let the reader reduce it in his thoughts to one-sixth, for it was not larger than an ordinary-sized cat.

A crocodile and the fleshless skeleton of the Ichthyosaurus remind us that reptiles still occupied an important place in the animal creation. A few insects, especially dragon-flies, fly about in the air. Ammonites float on the surface of the waves, and the terrible Plesiosaurus, like a gigantic swan, swims about in the sea. The circular reef of coral, the work of ancient polypi, foreshadow the atolls of the great ocean, for it was during the Jurassic period that the polypi of the ancient world were most active in the production of coral reefs and islets.

Louis Figuier, *Earth before the Deluge* (1863).

Figure 90. "Ideal View of the Earth
during the Period of the Lower
Oolite," from Louis Figuier's *Earth
before the Deluge* (1863).

Text 72

[Fig. 91] represents, after a sketch by M. Eudes Des-longchamps, the *Teleosaurus cadomensis*, carrying out of the sea a *Geoteuthis*, a kind of squid of the Oolitic epoch. This creature has the curious peculiarity of being cuirassed both on back and belly. In order to show this peculiarity, a living individual is repre-sented on the shore, and a dead one is floating on its back in shallow water, leaving the ventral cuirass exposed.

Behind *Teleosaurus cadomensis* in the engraving,

another Saurian, the *Hyleosaurus*, is represented, which we will meet again in the cretaceous epoch. We have here adopted the restoration which has been so ably executed by Mr. Waterhouse Hawkins, at the Crystal Palace, Sydenham.

Louis Figuier, *Earth before the Deluge*, fourth edition (1865).

NOTE. *Eugène Eudes-Deslongchamps* (1830–89) was a French paleontologist well known for his work on these Jurassic reptiles.

Figure 91. "The Teleosaur and the Hylaeosaur (Middle Oolite Period)," from the fourth edition of Louis Figuier's *Earth before the Deluge* (1865).

Text 73

[Fig. 92,] which represents a view of the earth during the upper Oolitic period, is intended principally to demonstrate the character of the vegetation of the Jurassic period. The *Sphenophyllum* among the tree-ferns are predominant in this vegetation; some *Pandanas*, a few *Zamites*, and many *Conifers*, but we perceive no palms. A coral islet rises out of the sea, having somewhat of the form of the *atolls* of Oceania, indicating the importance these formations assumed in the Jurassic period. The animals represented are the *Crocodileimus* of Jourdan, the *Ramphorynchus*, with the imprints which characterise its footsteps, and some of the invertebrated animals of the period, as the Asterias, Comatulas, Hemicidaris, Pteroceros. Aloft in the air floats the bird of Solenhofen, the *Archaeopteryx*, which has been reconstructed from the skeleton, with the exception of the head, which remains undiscovered.

Louis Figuier, *Earth before the Deluge*, fourth edition (1865).

Figure 92. "View of the Earth during the Upper Oolite Period," from the fourth edition of Louis Figuier's *Earth before the Deluge* (1865).

193

different reptile, the megalosaur, which appears to be based on Hawkins's model at the Crystal Palace.

The final scene from all the Secondary periods represents the Upper Cretaceous (fig. 94; text 75). Riou presents a loose adaptation of Kuwasseg's corresponding scene (fig. 52), much cruder in artistic terms but somewhat more informative scientifically. Reflecting a better understanding of the likely conditions of formation of the Chalk, this is not a storm-lashed mountainous coastline, but the low-lying shore of a vast and calm sea. Washed up onto the beach is the usual well-arranged selection of marine life; beyond is a similar selection of the trees of the period. The only major addition to Unger's scenario is the *Mosasaurus*, a giant marine lizard whose fossil skull had been a celebrated object since before the time of Cuvier.[6]

The first of Riou's scenes from the Tertiary periods depicts the Eocene (fig. 95; text 76). Figuier here adopts the term that Lyell had proposed thirty years before, for the period of formation of the strata around Paris. The scene is closely modeled on the one in Unger's book (fig. 53); borrowed elements include the pastoral landscape setting and details such as the watchful wader on a rocky perch. But as usual, the animals are made more prominent, and the mammals that Cuvier had first reconstructed are brought forward and highlighted in a patch of sun.

Riou's scene from the following period, Lyell's Miocene, makes the mammals even more clearly the main actors (fig. 96; text 77). In contrast to the scene in Unger's book (fig. 54), the dense vegetation is mere background and receives no comment in Figuier's text. Conversely, the mastodon lurking in the background of Kuwasseg's composition is brought forward and made as prominent as the nearby rhinoceros. But center stage is taken by the dinotherium, which sits peacefully by the edge of a pool,

Text 74

[Fig. 93,] which shows a struggle between an Iguanodon and a Megalosaurus, in the middle of a forest of the lower Cretaceous epoch, also enables us to convey some idea of the vegetation of the period. Here we note a vegetation at once exotic and temperate—that of the tropics, and a flora resembling our own. On the left we observe a group of trees, which resemble the dicotyledonous plants of our forests. The elegant *Credneria* is there, whose botanical place is still doubtful, for its fruit has not been found, although it is believed to have belonged to plants with two seed-leaves, or dicotyledonous, and the arborescent Amentaceae. An entire group of trees, composed of ferns and zamites, are in the background; in the extreme distance are some palms. We also recognize in the picture the alder, the wych-elm, the maple, and the walnut-tree, or at least species analogous to these.

Louis Figuier, *Earth before the Deluge* (1863).

Figure 93. "The Iguanodon and
the Megalosaur (Lower Cretaceous
Period)," from Louis Figuier's
Earth before the Deluge (1863).

Figure 94. "Ideal View of the Earth during the Upper Cretaceous Period," from Louis Figuier's *Earth before the Deluge* (1863).

Text 75

In [fig. 94] is represented an ideal view of the earth during the Upper Cretaceous period. In the sea swims the Mosasaurus: mollusks, zoophytes, and other animals proper to the period are seen on the shore. The vegetation seems to approach that of our days; it consists of ferns and pterophyllums, mingled with palms, willows and some dicotyledons of species analogous to those of our epoch. Some Algae, then very abundant, compose the vegetation of the sea-shore.

Louis Figuier, *Earth before the Deluge* (1863).

Text 76

In [fig. 95] we present an ideal landscape of the eocene period. We remark amongst its vegetation a mixture of the fossil species with others belonging to the present age. The alders, the wych-elms, and the cypresses, mingle with the *Flabellaria,* some palms of an extinct species. A great bird—a wader, the *Tantalus*—occupies the projecting point of a rock on the right: the turtle *Trionyx,* floats on the river, in the midst of Nymphaeas and Nenuphars and other aquatic plants. Whilst a herd of Palaeotheriums, Anoplotheriums and Xiphodons peacefully browse the grass of the wild meadows of this tranquil oasis.

Louis Figuier, *Earth before the Deluge* (1863).

Figure 95. "Ideal Landscape of the Eocene Period," from Louis Figuier's *Earth before the Deluge* (1863).

197

in exactly the same pose as in its first reconstruction by Kaup nearly thirty years earlier (fig. 30). One newcomer, less conspicuous but equally important, reflects more recent discoveries: perched in a tree is a *Dryopithecus*, a primate that had first pushed the record of manlike mammals back into relatively deep time.

In Figuier's first edition, the next two scenes depict the Pliocene, the third of the Tertiary periods that Lyell had defined. Here Figuier and Riou were on their own, with no model from Unger to inspire them. There are two scenes because, for the first time, Figuier makes a geographical distinction between the Old World and the New. The European Pliocene landscape (fig. 97; text 78) combines a pastoral foreground, to provide a plausible habitat for a varied mammalian fauna, with a background of volcanos, an allusion to the volcanic activity known from this period in several parts of Europe. Although some of the animals, such as the straight-tusked mastodon, are reconstructed from fairly complete fossil evidence, Figuier concedes that others, such as the horse, are simply drawn from living animals on the basis of much more fragmentary fossil remains. (Kaup's artists had earlier cut corners in the same way.)

The scene of American life in the Pliocene (not reproduced here) was designed to portray the spectacular mammals that had been found as fossils in the New World, particularly in South America; the landscape is closely similar to the European scene. However, after Figuier's book was published, he must have been informed, or persuaded, that these mammals were not Pliocene in age, but more recent. By the fourth edition, therefore, this scene had been withdrawn, and its denizens transferred to a newly composed picture of South America in the Pleistocene.

In fact, rather than using the term Pleistocene (the fourth and last of Lyell's divisions of the Tertiary), Figuier

Text 77
An ideal landscape of the miocene period, which is given in [fig. 96], represents the Dinotherium lying in the marshy grass, the rhinoceros, the Mastodon, and an ape of great size, the *Dryopithecus*, suspended from the branches of a tree.

Louis Figuier, *Earth before the Deluge* (1863).

Figure 96. "Ideal View of the
Earth during the Miocene Period,"
from Louis Figuier's *Earth before
the Deluge* (1863).

Figure 97. "Ideal View of a European Landscape during the Pliocene Period," from Louis Figuier's *Earth before the Deluge* (1863).

Text 78

In [fig. 97] an ideal landscape of the pliocene period is given under European latitudes. At the bottom of the picture, a mountain recently thrown up reminds us that the period was one of frequent convulsions, in which the soil was disturbed and overthrown, and mountains and mountain ranges made their appearance. The vegetation is nearly identical with that of our days. We see assembled in the foreground the more important creations of the period—the fossil species have been restored from the remains which have been found, others from their descendants where they still exist.

Louis Figuier, *Earth before the Deluge* (1863).

Text 79

We have assembled in [fig. 98] the large edentates which lived exclusively in America during the Quaternary epoch: the Glyptodon, the Megatherium and the Mylodon; to which is added the Mastodon. A little macaque monkey, the Oreopithecus, which had already appeared in the Miocene period [see fig. 96] hangs from the trees of this landscape, in which the vegetation is similar to that of the present-day regions of equatorial America.

Louis Figuier, *Earth before the Deluge*, fourth edition (1865).

Figure 98. "Ideal View of a [South] American Landscape during the Quaternary Epoch," from the sixth edition of Louis Figuier's *Earth before the Deluge* (1867).

adopts an almost synonymous term for this most recent portion of geological time, the Quaternary epoch. By the fourth edition, as a result of the alteration just mentioned, it was the Quaternary, not the Pliocene, that had two scenes distinguished geographically. Riou was perhaps temporarily unavailable; in any case, the American Quaternary was at first represented by an unsigned scene (not reproduced here) of distinctly inferior quality. However, by the sixth edition (1867) Riou had returned with a much more impressive scene (fig. 98; text 79). The largest mammal, curiously indistinct in form, is a megatherium, whose skeleton had first drawn Cuvier's attention to fossil anatomy almost seventy years earlier (fig. 12). A *Mylodon*, or giant sloth, reaches up into a tree; the mastodon makes another appearance; and spotlit in the foreground are a giant armadillo, the *Glyptodon*, and a small primate, the *Oreopithecus*.

Meanwhile, in northern Europe, life in the Quaternary epoch was much harder (fig. 99; text 80). Once again, Riou's scene is loosely based on Kuwasseg's (fig. 55), though inferior in artistic terms. Cave bears enjoying a mammoth dinner again occupy the foreground, though here they are challenged by one of Buckland's cave hyenas; in the middle ground, less convincing ecologically than in Unger's scene, is a much wider selection of Pleistocene mammals; and in the background are ice-capped mountains that presage the impending glacial period or Ice Age. As in Unger's sequence, this is still a prehuman world.

Unger had called Kuwasseg's scene a picture of the Diluvial period. But he used that term in a technical sense, and he clearly did not believe there had been any but strictly local and small-scale deluges, caused, for example, by the bursting of glacial lakes. Figuier, on the other hand, exploited the popular sense of "the antediluvian world" by choosing to entitle his book *Earth before the Deluge*.

Text 80
In [fig. 99] an attempt is made to represent the appearance of Europe during the epoch we have under consideration. The bear is seated at the mouth of its den—the cavern, thus reminding us of the origin of its name of *Ursus spelaeus*, where it gnaws the bones of the elephant. Above the cavern the *Hyaena spelaea* looks out with savage eye for the moment when it will be prudent to dispute possession of these remains with its formidable rival. The great wood-stag, with other great animals of the epoch, occupies the farthest shore of a small lake, where some small hills rise out of a valley crowned with the trees and shrubs of the period. Mountains recently upheaved rise on the distant horizon, covered with a mantle of frozen snow, reminding us that the glacial period is approaching; which, by freezing part of the globe in an unanticipated way, would cause the rapid extirpation of the mammoths and *Rhinoceros tichorhinus*, effacing their races from the surface of the earth.

Louis Figuier, *Earth before the Deluge* (1863).

Figure 99. "Ideal View of the Earth during the Quaternary Epoch (Europe)," from Louis Figuier's *Earth before the Deluge* (1863).

At some point he would have to cash that promissory note, by describing something more than a little local difficulty. Yet he was well aware that geologists had long been skeptical about the reality of any widespread, let alone global, catastrophe of the kind that Buckland had postulated some forty years earlier. He resolved that dilemma by making a move that was sanctioned by at least part of the geological community in the 1860s. He distinguished *two* Deluges of contrasting character. Neither was global, though both had wide effects; but most important, one was prehuman, the other within human history (text 81). .

Figuier termed the earlier event the "Deluge of northern Europe," and Riou drew an appropriately violent scene to illustrate it (fig. 100). Its inferred cause was the elevation of the mountains of Scandinavia; as geologists still commonly assumed, this was taken to be due to a sudden buckling of part of the earth's crust. The effect of this putative event was to flood northern Europe with an equally sudden, violent but transient rush of waters bearing floating icebergs. This theory served to explain the distribution of the very peculiar deposits that were termed Diluvium or Drift. For example, large areas of Scandinavia, the north German plain, and much of lowland Britain were covered with Boulder Clay or Till, containing "erratic boulders" of rocks that could only be matched with sources tens or hundreds of miles to the north.

Agassiz's theory of an Ice Age had made headway only in a modest form, as an explanation of the evidence for the formerly greater extent of glaciers such as those in the Alps. The Diluvial features seemed to present an altogether different problem, and many geologists doubted whether they were due to ice-sheets vastly greater than any European glaciers. So the European Deluge that Riou portrayed was huge and violent in character (the surviving conifers indicate the scale), and near-glacial in

Text 81

There is very distinct evidence of two deluges succeeding each other in our hemisphere during the quaternary epoch. One we shall distinguish as the two *European deluges;* the other as the *Asiatic.* The two European deluges were anterior to the appearance of man; the Asiatic deluge is posterior to that event: the human race, then, in the early days of its existence certainly suffered from a great deluge. In the present chapter we confine ourselves to the two cataclysms which overwhelmed Europe in the quaternary epoch.

The first occurred in the north of Europe, and it was produced by the upheaval of the mountains of Norway. Commencing in Scandinavia, the wave spread and carried its ravages into those regions which now constitute Sweden, Norway, European Russia, and the north of Germany, sweeping before it all the loose soil on the surface, and covering the whole of Scandinavia—all the plains and valleys of Northern Europe—with a mantle of shifting soil. As the regions in the midst of which this great mountainous upheaval occurred—as the seas surrounding these vast spaces were partly frozen and covered with ice, from their elevation and neighbourhood to the pole—the wave which swept these countries carried along with it enormous masses of ice: the collision of these several solid blocks of congealed water would only contribute to increase the extent and intensity of the ravages occasioned by this violent cataclysm, which is represented in [fig. 100].

Louis Figuier, *Earth before the Deluge* (1863).

Figure 100. "Deluge of Northern
Europe," from Louis Figuier's *Earth
before the Deluge* (1863).

climate. But it was confined to northern Europe, and it predated the appearance of man.

The creation of man evokes Figuier's best purple prose (text 82). That great event is described in conventionally theistic terms; the physical nature of the event is left conveniently vague. More important is the way it is described as the culmination of all the vast history that the book has hitherto described. Just as for Unger, the appearance of man is the goal toward which the whole drama has been designed.

This much was entirely conventional for the mid nineteenth century. It is more surprising, however, to find Figuier in the 1860s instructing Riou to portray the appearance of man in a way that makes no allusion to the current debates about the origin and antiquity of the human species. The scene that appeared in the early editions of Figuier's book (fig. 101) is similar in character to Kuwasseg's final scene for Unger (fig. 56). The first human family is highlighted in the foreground of a landscape as Arcadian as Kuwasseg's. The man again carries only a staff, to help him herd the animals that graze nearby, awaiting domestication. Only a distant deer suggests the wilder nature that may need to be hunted rather than herded. The larger and more hostile creatures that have dominated earlier scenes have vanished, thanks to the ravages of the glacial period and the prehuman Deluge. Though the man's modest covering might allude to the fig leaves of the story of Eden, this is no world of thorns and hard labor; there is no sweat on this Adam's brow. He and his spouse are also, as in Kuwasseg's scene, unmistakably white and European, modern and noble human beings. Indeed, in his commentary (text 82) Figuier forcefully rejects the low and horrible vision of the earliest human beings that his fellow-Parisian Boitard had presented only two years earlier (fig. 77; text 59).

Unluckily for Figuier, the debate about human antiq-

Text 82

It was only after the glacial period, when the earth had resumed its normal temperature, that man was created. Whence came he?

He came whence the first blade of grass which grew upon the burning rocks of the Silurian seas came; from whence came the different races of animals which have from time to time replaced each other upon the globe, gradually rising in the scale of perfection. He emanated from the will of the Author of the worlds which constitute the universe.

. . . . .

At the close of the tertiary epoch, the continents and the seas have assumed the respective limits which they now present. The disturbances of the soil, the fractures of the crust and volcanic eruptions and earthquakes of which they are the consequences, only occurred now at rare intervals, occasioning only local and restricted disasters. The rivers and their affluents flowed between tranquil banks; the animated creation was that of our own days. An abundant vegetation, diversified by the existence of a climate which has now been acquired, embellishes the earth. A multitude of animals inhabit the waters, the dry land and the air. Nevertheless, creation has not yet achieved its greatest work; a being capable of comprehending these marvels and of admiring the sublime work—a soul is wanting to adore and give thanks to the Creator.

God created man.

. . . . .

Volumes have been written upon the question of the unity of the human race; that is, whether there were many centres of the creation of man, or if the parent of our race was the Adam of Scripture. We think, with many naturalists, that the stock of humanity is unique, and that the several races of negroes, black and yellow, are only the result of climate upon organism. We consider that the human race appeared for the first time with a divine mystery which is eternally impenetrable to us as to the

Figure 101. "Appearance of Man,"
from Louis Figuier's *Earth before the
Deluge* (1863).

uity took a new and decisive turn just as his book was being published. Boucher de Perthes found a human jaw in the Somme gravels that he had long been investigating. Added to the earlier evidence of flint axes and other apparently human artifacts in the same deposits, this seemed to swing the argument in favor of those who claimed that the first human beings had indeed coexisted with the extinct fauna that Cuvier had resurrected half a century earlier.[7]

Figuier added a defensive footnote to the next edition (text 83), claiming that he had placed the appearance of man within the Quaternary epoch even *before* Boucher's latest find. But in his sixth edition (1867), he bowed to the inevitable, and got Riou to draw a completely new scene (fig. 102) to replace the Arcadian original. The design recalls Boitard's picture of cave-dwelling primitive men (fig. 78), though of course the borrowing remains unacknowledged. Riou's scene is superior in quality, both artistically and scientifically; but still, in both scenes, the human beings outside their cave confront hostile nature across a defensive gully that symbolically divides the human world from the nonhuman.

Figuier left his text unaltered (apart from the footnote), so that his denial of the cave-dwelling habit of the first human beings is contradicted by Riou's new scene. The publishers of some later editions in English, both British and American, tried to have the best of both worlds: the first version of the scene was retained as the frontispiece of the book—where it might reassure conservatively minded purchasers—while the second version was inserted in its proper position near the end.

Riou's human beings are still as modern in physiognomy, and as white and European, as those in his original scene; there is no suggestion here of the bestiality that Boitard had witnessed on his magical time-travel (text 59). In place of the primal nuclear family, however, Riou

mode of creation. In the rich plains of Asia, on the smiling banks of the Euphrates, as the traditions of the most ancient races teach us—in the midst of this rich and vigorous soil, under the brilliant climate and the radiant sky of Asia, in the shade of its luxuriant masses of verdure and its mild and perfumed atmosphere, man loves to represent to himself the father of his race issuing from the hand of his Creator.

We are, thus, far from sharing the opinion of those naturalists, if there be any such, who represent man at the beginning of the existence of his species as a sort of ape, of hideous face, degraded mien, and covered with hair, inhabiting caves like the bears and lions, and participating in instincts at once brutal and ferocious. There is no doubt the primitive man had to pass through a period in which he had to contend for his existence with ferocious beasts, to live in the woods like the apes, and in the savannahs, where Providence had thrown him. But this period of probation came to an end, and man, an eminently social being, promptly combined in groups, animated by the same interests and the same desires, and alike destined to triumph over the elements, and to subdue to their rule the other inhabitants of the earth.

Louis Figuier, *Earth before the Deluge* (1863).

Text 83

The discovery of a human jaw, made by M. Boucher de Perthes in April 1863 in the Quaternary deposits of Moulin-Quignon near Abbeville, added to the mass of facts previously known, which revealed the existence in those same deposits of traces of human industry such as flint axes, and remains of hearths and pottery, has established in a striking manner that man existed during the Quaternary period and before the Asiatic Deluge. It was in the light of facts long known that, in the early editions of this work, the birth of the human species was placed during the Quaternary epoch. The discovery of fossil man,

Figure 102. "Appearance of Man,"
revised version, from the sixth edition of Louis Figuier's *Earth before
the Deluge* (1867).

now depicts a larger tribal group, with a sharp functional differentiation of the sexes that would have reassured his bourgeois readers. The Edenic man's pastoral staff is replaced by a stout flint-headed axe identical to Boitard's, and his fig leaves are replaced by animal skins like those that covered Boitard's simian creatures. However, the Pleistocene mammals that menace the human beings from the other side of the gully are far more convincing than Boitard's: notably cave hyenas and bears, with mammoth and rhinoceros standing further back. Only a solitary deer, as in the original version, suggests a nature that is wild but not hostile; only an implausibly thoroughbred horse suggests a nature that man may be able to subdue to his will. The clearly subtropical vegetation, which is here combined incongruously with the animals of cold northern Europe, alludes to the traditionally southern and Asiatic birthplace of the human species, to which Figuier's text explicitly refers.

Figuier's dramatic change of scene, right in the middle of a series of successful editions, neatly encapsulates the historical moment at which the human species was brought fully into the visual panorama of the history of life on earth. With this revised version of "the appearance of man," the human culmination of that history is represented for the first time in just the same pictorial style as any scene from the earlier periods of deep time.

With the appearance of man, in either version of the scene, Figuier might well have followed Unger in regarding his narrative task as complete. But one item on his agenda remained. The Deluge of his title had been expounded so far as an event that was clearly *not* the one that many of his readers would have expected. As a definitely prehuman event, it could not be the Deluge that was recorded both in Genesis and—it was widely believed—in other ancient human records too. So Figuier got Riou to round off his great sequence of scenes with

made in 1863 by M. Boucher de Perthes, has thus only confirmed the views already expressed in this book in 1862.

Louis Figuier, *Earth before the Deluge*, note added in fourth edition (1865).

NOTE. The reference to 1862 is either a convenient misremembering of the publication date of the first edition or (more charitably) refers to the time Figuier was composing his original text; in either case, it made his assertion safely prior to Boucher's 1863 find.

one that would represent that later and human catastrophe. This was the second of the two Deluges to which he had earlier referred, and he distinguished it as the "Asiatic Deluge" (fig. 103; text 84).

This final scene is unmistakably inspired by John Martin's famous engraving of the Flood (fig. 10), now forty years old but as popular as ever, in France as much as in England. There are the same dramatic contrasts of darkness and light, the same vast vortices of swirling waters, the same torrential rains and flashes of lightning. The human presence, however, is indicated not by crowds of men and women striking poses of despair, but only by a distant view of an imaginary antediluvial city. But even this detail is derived from Martin's work: not from his picture of the Flood but from the fantastic architecture of "The Fall of Babylon" and other such designs. The threat to the natural world is likewise reduced here to a couple of trumpeting elephants and a few cedars about to be swept away by the rising waters.

For a scene that stands in the long artistic tradition of representing Noah's Flood, the absence of the Ark itself is at first glance surprising. Even the faint and distant vessel of Martin's design has vanished. But Figuier's text shows how far he had diverged from any scriptural literalism. For him, as for many others at this period, the "Asiatic Deluge" was just that: an event confined to Mesopotamia. For Figuier its cause was entirely natural; like the earlier European Deluge, it was due to the sudden upheaval of distant mountains, in this case those of the Caucasus, and specifically the area around the volcanic mountain Ararat, the traditional landing site of the Ark. But that natural cause is of course perfectly compatible with a providential purpose. Figuier can therefore conclude that, having interpreted the biblical Deluge as a local event—which was no hermeneutic innovation—its religious significance is unaffected.

211

That conclusion, like the conventional theistic language Figuier used throughout his interpretation of the history of life, doubtless contributed to the acceptability and popularity of the work. Its rapid sequence of French editions has already been mentioned. By the fourth edition (1865), only two years after the first, its sales had reached 25,000 copies, an astounding figure at this period for a serious work of nonfiction in French. It was the fourth edition that was translated—by a member of the Geological Society of London—to form the first English edition (1865). That in turn was revised two years later, in time to catch the new version of "The Appearance of Man" from the sixth French edition (1867).[8] The English translation was also published in Philadelphia for the North American market, and an edition in Spanish was published in Mexico for the Latin American market. There were even editions in Danish and probably other languages too.

One apparently glaring gap in this summary of worldwide distribution is the lack of an edition in German, which by this time had overtaken French as the major international language of science. But this is easily explained by the almost simultaneous appearance of a popular book in German with an almost identical title. *Before the Deluge!* (*Vor der Sündfluth!*, 1866) was by the German geologist Oskar Friedrich von Fraas (1824–97), the geological curator at the natural history museum in Stuttgart. According to the prospectus for this book, it was the success of Figuier's book in France that prompted Fraas to write a similar work. But it cannot have been a case of outright competition, for Figuier and his publisher allowed Fraas to reissue most of Riou's scenes.[9] Fraas's vision of the deep past, like Figuier's, is explicitly framed in terms of the divine purpose embodied in the ascent of life toward Man. However, Fraas seems to have balked at reproducing Riou's Edenic scene of "the

Text 84

The opinion which places the creation of man on the banks of the Euphrates in Central Asia is confirmed by an event of the highest importance in the history of humanity, and by a crowd of concordant traditions, preserved by different races of men, all tending to confirm it. We speak of the Asiatic deluge.

The Asiatic deluge, of which sacred history has transmitted to us the few particulars we know, was the result of the upheaval of a part of the long chain of mountains which diverge from the Caucasus. The earth opening by one of these fissures left in its crust in course of cooling, an eruption of volcanic matter escaped through the crater. Masses of watery vapour or steam accompanied the lava discharged from the interior of the globe, which, being dissipated in clouds and condensing, descended in torrents of rain, and the plains were drowned under the volcanic mud. The inundation of the plains in an extensive radius was the momentous result of this upheaval, and the formation of the volcanic cone of Mount Ararat, with the vast plateau on which it rests, altogether 17,323 feet above the sea, the permanent result. The event is graphically detailed in the seventh chapter of Genesis.

. . . . .

Nothing occurs, therefore, in the description given by Moses to hinder us from seeing in the Asiatic deluge a means adopted by God to punish the human race still in its infancy. It seems to establish the countries lying at the foot of the Caucasus as the cradle of the human race, and it seems to establish also the upheaval of a chain of mountains, preceded by an eruption of volcanic mud, which drowned vast territories in these regions, consisting of plains of great extent. Of this deluge many races besides the Jews have preserved the tradition.

. . . Thus the Biblical deluge [fig. 103] is confirmed in many respects, but it was local, like all phenomena of the kind, and was the consequence of the upheaval of the mountains of western Asia.

Louis Figuier, *Earth before the Deluge* (1863).

Figure 103. "The Asiatic Deluge,"
from Louis Figuier's *Earth before the
Deluge* (1863).

appearance of man" (fig. 101), perhaps because it could not claim to be based on any scientific evidence at all. But like Figuier again, Fraas regards the Deluge of his title as a decisive boundary between prehuman and human, between the "primaeval world" and the present world (text 85), and he even uses Riou's Martinesque scene of the "Asiatic Deluge" (fig. 103) as his frontispiece.[10]

Fraas's text follows the pattern of Figuier's, in tracing the temporal development of the earth and its life from the most distant past to its culmination in the present, human world. But for promotional purposes, Fraas—or at least his Stuttgart publisher—had no compunction about compressing this complex story into a *single* world "before the Deluge." The design on the front cover of each paperback installment (fig. 104) is clearly borrowed from the scenes in Figuier's book, particularly the spouting ichthyosaur, and the iguanodon and megalosaur locked in mortal combat (figs. 88, 89). But rather like Milner's earlier decorative design (fig. 41), it mixes trilobites, ichthyosaurs, and mammoths in a single scene. This evidently remained an acceptable device for catching the attention of the prospective purchaser; the text of the book, with Riou's scenes, could then be expected to enlarge that popular notion of an undifferentiated "antediluvian" world into a richer conception of a highly diversified past.

As a work that was both instructive and entertaining, Figuier's *Earth before the Deluge* doubtless appealed widely to middle-class adults as a suitable Christmas present for their children. In his preface, Figuier had argued that popular books on natural history should replace traditional fairy tales and myths as the imaginative diet of the young (text 86). That earnest proposal was mocked in *Punch*, in its issue for Boxing Day 1868—the day after Christmas, when Victorian children were free to enjoy their presents—in a way that illustrates yet again the

Text 85

Deep in human nature lies the craving for a detailed understanding of the emergence and development of the globe, and consequently the irrepressible urge to fathom the mysterious processes that constituted the primaeval period [*Urzeit*] of our planet. Although this urge was often suppressed through the centuries, it has broken through more and more, and nowadays science penetrates ever deeper into the consciousness of the people.

Our times demand that the educated person, who is to be entrusted with the history of *his own kind*, should also be familiar with the *history of the planet* on which we live. To each should be given an insight into the mysterious life and activity on earth, from the monad whose world is a drop of water, upwards to the most complete creature, Man!

The picture of creation is indeed set out so in such beautiful colours in *sacred scripture* that even today, as for thousands of years, it still holds our senses captive in the power of its original freshness. Yet in line with the progress of knowledge, that picture demands an explanation, which in no way obscures the pure account of the history of creation in scripture—still less the glory of God, who discloses himself in his eternal plan of creation—but puts it in a clearer light.

Unfortunately the science of earth-history has often come into disrepute, for many *fictions* have been put forward as truth, and the products of fantasy mistaken for the results of earnest research and calm observation. In the present work, *truth*, as far as it is known, has been the guiding thread before all else, and the author has endeavoured in this spirit to make the latest results of scholarly research trustworthy and intelligible to educated people. The reader is introduced to the inner relation between the planet and all living things, and is shown the organic development that has evolved in the course of aeons from crystal to Man.

In three ages of the world the simple bodies of

Figure 104. Cover design for the paperback installments of Oscar Fraas's *Before the Flood!* (1866), which incorporated most of Riou's original illustrations for Louis Figuier's *Earth before the Deluge*.

imaginative plasticity of scenes from deep time (fig. 105). The cartoon by George du Maurier (1834–96), John Leech's successor with *Punch*, shows the fearful creatures of John Martin's imagination (figs. 35, 36) transmuted into the unnervingly Freudian monster of a little boy's nightmare, perhaps induced by a surfeit of Christmas pudding. A huge woolly mammoth pursues the terrified child down a snowy suburban street, observed by a stereotypically unsurprised English policeman. The heavy dreamlike sensation of the boy's snow-caked feet is matched by the fearsome metamorphosis of the mammoth. Its trunk has developed a terrible second head with open jaws and savage teeth, which are about to devour their youthful victim. That myth-making use of Figuier's work is a striking early example of the ways in which fearsome dinosaurs have functioned ever since, serving psychological and perhaps cultural purposes of unfathomable depth. The scary monsters of the prehuman world are alive and well, and lying in wait in every natural history museum.

The sequence of pictures that Riou drew for Figuier established the genre of scenes from deep time, not only for the scientists whose research had provided the materials for the reconstructions, but also for a vast literate public spread around the Western world. While borrowing from earlier models, particularly the scenes that Kuwasseg had drawn for Unger, Figuier's book illustrated the long history of life on earth in greater and more specific detail than ever before, and made it far more widely accessible than Unger's *de luxe* work had ever been. However much the component features may

the earth have evolved into ever more diverse organic beings, until the great work was complete and Man was created. But only with the *Deluge*, which gives this work its title, did the earth's final preparation for Man take place; the Deluge is the boundary stone between the present world and the primaeval world. With it begins the present state of the surface of the planet, and the development of the human race, in which the divine creative plan has reached its peak.

"Prospectus" for Oscar Fraas, *Before the Deluge!* (1866).

Text 86

I am about to maintain what may be thought a strange thesis.

I assert that the first books placed in the hands of the young, when they have mastered the first steps to knowledge and can read, should be on Natural History; that in place of awakening the faculties of the youthful mind to admiration, by the fables of Aesop or Fontaine, by the fairy tales of 'Puss in Boots,' 'Jack the Giant Killer,' 'Cinderella,' 'Beauty and the Beast'—or even 'Aladdin and the Wonderful Lamp,' and such purely imaginative productions, it would be better to direct their admiring attention to the simple spectacles of nature—to the structure of a tree, the composition of a flower, the organs of animals, the perfection of the crystalline form in minerals, above all, to the history of the world—our habitation; the arrangement of its stratification, and the story of its birth, as related by the remains of its many revolutions to be gathered from the rocks beneath our feet.

Many readers will protest against this proposition. Is it not the fact, they will say, that fairy tales, fables, and the legends of mythology, have always been the first intellectual food offered to the young? Is not that the natural means of amusing them, as

Figure 105. "A Little Christmas Dream": a cartoon by George du Maurier in *Punch* (1868).

A LITTLE CHRISTMAS DREAM.

MR. L. FIGUIER, IN THE THESIS WHICH PRECEDES HIS INTERESTING WORK ON THE WORLD BEFORE THE FLOOD, CONDEMNS THE PRACTICE OF AWAKENING THE YOUTHFUL MIND TO ADMIRATION BY MEANS OF FABLES AND FAIRY TALES, AND RECOMMENDS, IN LIEU THEREOF, THE STUDY OF THE NATURAL HISTORY OF THE WORLD IN WHICH WE LIVE. FIRED BY THIS ADVICE, WE HAVE TRIED THE EXPERIMENT ON OUR ELDEST, AN IMAGINATIVE BOY OF SIX. WE HAVE CUT OFF HIS "CINDERELLA" AND HIS "PUSS IN BOOTS," AND INTRODUCED HIM TO SOME OF THE MORE PEACEFUL FAUNA OF THE PREADAMITE WORLD, AS THEY APPEAR RESTORED IN MR. FIGUIER'S BOOK.

THE POOR BOY HAS NOT HAD A DECENT NIGHT'S REST EVER SINCE!

217

have been modified or even transformed by subsequent research, Riou's illustrations have set the pictorial precedents, and established the visual language, for all later scenes from deep time, not least those in our modern museums and television programs.

some relaxation from more severe study?

And, they will add, society has been none the worse!

It is here I would claim the reader's attention. I think, on the contrary, that many of the evils of society may be traced to this very cause. It is because we cherish this dangerous aliment, that the living generation includes so much that is false, so many weak and irresolute minds, given to credulity, inclined to mysticism—proselytes, in advance, to chimerical conceptions and to every extravagant system.

Intelligence is scarcely awakened, when we do our best to destroy it by our training. Our very first step in this path of folly is to teach the impossible and absurd. We crush, so to speak, good sense in the eggshell, when we concentrate the ideas of the young on conceptions at once dreamy and opposed to fact and reason; introducing them into a fantastic world, in which are jumbled together gods, demigods, and pagan heroes, mingling with fairies, goblins, and sylphs; spirits—good and bad—enchanters, magicians, devils, devilkins, and demons; and all this while no doubts seem to be entertained of the danger likely to arise from this constant presence of ideas so subversive of common sense.

Introductory "Thesis" to Louis Figuier's *Earth before the Deluge* (1863).

7

Making Sense of It All

Riou's sequence of pictures for Figuier established the genre of scenes from deep time throughout the Western world. Later nineteenth-century versions of such scenes were clearly modeled on Riou's, and their influence has continued, however indirectly and unrecognized, throughout the twentieth century.[1] They themselves were modeled on Kuwasseg's earlier sequence for Unger, and on other precedents that stretch back to De la Beche's innovative scene of "a more ancient Dorset." However, even that prototype was based in turn on animal reconstructions that go back to Cuvier at the start of the nineteenth century, and on a still older tradition of natural history illustration; while the idea of portraying a sequence of scenes finds its precedents in the equally ancient tradition of biblical illustration. Having traced the origins, emergence, and consolidation of the genre of scenes from deep time, we are in a position to reflect on some more general issues that the narrative raises.

Riou's pictures for Figuier embody much that modern scientists would still endorse as reliable inferences from the surviving relics of the deep past. Where modern scenes differ from Riou's, it is not unreasonable to attribute the changes, at least in part, to the subsequent growth of scientific knowledge. In a science such as paleontology—and scenes from deep time have always been dominated by portrayals of the living world inferred from fossils—there is indeed an unmistakable element of progressive improvement in the development of knowledge. With the luxury of hindsight, scientists may judge that their predecessors were sometimes gravely in error in matters of theory; but no one, scientist or not, can reasonably deny the element of plain cumulation that is reflected most weightily in the ever-growing stacks of specimens stored in our museums. Quite simply, scientists of later periods have the benefit of more (and often better) specimens—in modern jargon, of a larger database—than were available to their predecessors. (When, rarely, the reverse is the case, because specimens have been lost or destroyed, it is rightly the occasion of loud lamentation!).

This obvious point needs to be stressed, because a simple interpretation would attribute the emergence of the genre of scenes from deep time to *nothing more* than the cumulation of material remains that made such reconstructions possible at a certain period of history, when they had not been possible before. Likewise, the changes that can be traced in successive attempts to portray the *same* scene from deep time could be attributed simply to the discovery of more and better specimens that are relevant to that reconstruction. There is of course much truth in this argument. Modern portrayals of the bipedal iguanodons, for example, are strikingly different from the giant lizards that Martin drew for Mantell (fig. 35), and from the rhinoceroslike monsters that Hawkins later constructed to Owen's instructions at the Crystal Palace (fig. 60); and modern reconstructions of the ichthyosaurs make them even more convincingly dolphinlike than the creatures that Conybeare first assembled.[2] In both cases, it is easy to argue that the modifications have been directly due to the later discovery of far more complete specimens than those that the scientists and artists of the mid nineteenth century had at their disposal. More generally, it is easy to argue that one reason why there were no true scenes from deep time before the nineteenth century is that until that period no extinct organisms had been reconstructed that could have figured in any such scene and made it significantly different from a scene of present-day life.

This kind of rational interpretation of the historical story could be restated in terms of a "cascade" of representations that allows progressively bolder—yet still well-founded—reconstructions of the unobservable prehuman past to be made.[3] One could argue that for any given organism the reconstruction necessarily progresses from the observed to the inferred, from the specific and contingent to the general and idealized. Taking vertebrate animals as an example—and they have always been the pacesetters in the construction of scenes from deep time—the cascade would begin with particular specimens of fossil bones, which may be identified and assembled as a partial skeleton of a particular individual. It would next proceed to the reconstruction of a complete skeleton representative of the species, based generally on the remains of many individuals. It would then extend to inferences about the animal's unpreserved muscles and other "soft parts," based partly on anatomical analogy with related living forms; and so to the reconstruction of a generalized complete individual body. The next stage would lead to inferences about the animal's dynamic mode of life and habits, based partly on functional analysis of its anatomy and on physiological analogy with related living forms. Finally, there could be an imaginative reconstruction of a complete scene, which would assemble and integrate similar reconstructions for many organisms that coexisted at a specific period of geohistory; this synthesis would be based partly on ecological analogy with existing habitats and would be set against a back-drop based on similar inferences from traces of the in-organic environment.

In part, the story that has been told in earlier chapters of this book conforms quite well to that idealized model. Cuvier's newly rigorous comparative anatomy allowed him to take the process down the first steps of the cas-cade, as far as the assembly of complete reconstructed bodies for some of his fossil mammals (fig. 15). Similar reconstructions of the Jurassic reptiles, supplemented by concrete evidence of their eating habits, later provided the necessary materials for De la Beche's first true scene from deep time, in which several organisms were shown in ecological interaction (that is, eating and being eaten!) within a generalized portrayal of the environment of "a more ancient Dorset" (fig. 19). A similar cascade lay tac-itly or explicitly behind many later scenes; Figuier, for ex-ample, made a didactic point of repeatedly illustrating the successive stages of inference that gave Riou's scenes their scientific validity.

The model of an inferential cascade can even be made to explain the greatest "errors" in the story. For example, the early reconstructions of the iguanodon, just men-tioned, can be faulted for having been based—in con-trast to all the other reconstructions that had been made by then—on nothing more than a few teeth and isolated bones. The early assumption that pterodactyles (or ptero-saurs, to call them now by their modern name) were ca-pable of powered flight can be faulted likewise for being based on an inadequate understanding of the physiology of living birds and bats.

However, this kind of rational reconstruction implies that scenes from deep time simply represent in visual terms the process of scientific *discovery:* as if the world of prehuman nature disclosed itself in terms that were only temporarily ambiguous, until better specimens allowed an asymptotic approach to a uniquely *true* representation of that vanished reality. But the examples analyzed in the earlier chapters of this book must surely suggest that it is more illuminating to regard scenes from deep time as human *constructions:* not unconstrained by the natural evidence available at a given period, but certainly using that evidence in a representation that has many other inputs beside the fossil bones and shells themselves. To interpret these scenes as a simple record of discovery may be congenial to some scientists, and to others to whom the notion of science as a human construction is anathema. But to evade the category of construction is to leave unanswered, indeed unasked, the most important questions that are posed by the history of the establishment of scenes from deep time as a mode of representing the prehuman world. The rest of this chapter will raise some of these questions and suggest at least some possible answers.

It is not, however, my intention to impose a Procrustean "theoretical framework" of interpretation. This would be as unfaithful to the marvelous multivocality of these materials as is the modern style of museum display that nags or bullies the visitor into viewing the exhibits in only one approved order, imposing a single—scientifically or politically "correct"—meaning on them. In contrast, I hope—in a more liberal, indeed, democratic mode—to leave readers free to "read" these scenes how they will, while suggesting some interpretative directions that I find persuasive, or at least intriguing and worth further exploration.

First, however, it is useful to summarize in visual form the historical sequence that has been traced in earlier

Figure 106. Summary of the early history of scenes from deep time. Square symbols denote pictorial scenes; diamonds denote reconstructions of individual organisms, not set in a landscape scene; one historically important set of *verbal* scenes is shown with small circles. Since the visual impact of pictorial scenes is so closely related to their absolute size, that dimension is indicated by the size of the symbols (denoting those of quarto size or larger, those that fit on an octavo page, and those that are even smaller). Solid black symbols denote scenes or reconstructions that were innovative, either because they depicted a period of earth history or an organism not previously represented or because they did so in a novel way; open symbols denote scenes that derived most of their form and content from some earlier precedent. The dashed and arrowed lines that connect some symbols likewise denote pictorial borrowing by later authors (or their artists) from earlier scenes, where such borrowing was either acknowledged explicitly or can be inferred plausibly from internal evidence. (Scenes are placed on the diagram at their dates of publication, except that Unger's original sequence is shown at its date of completion.)

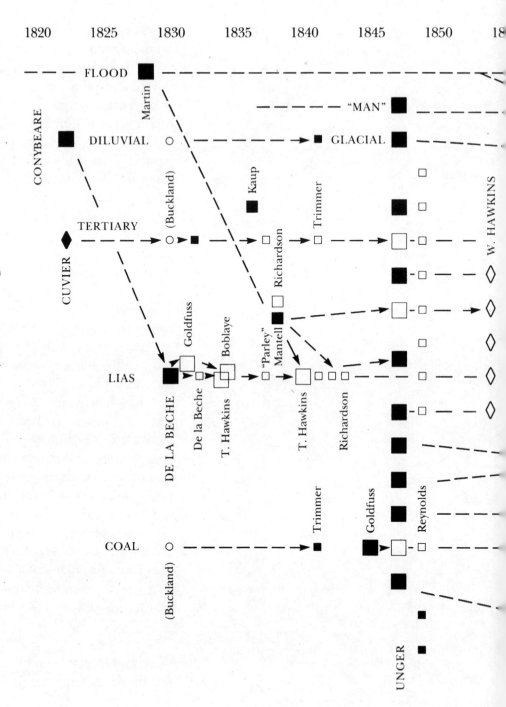

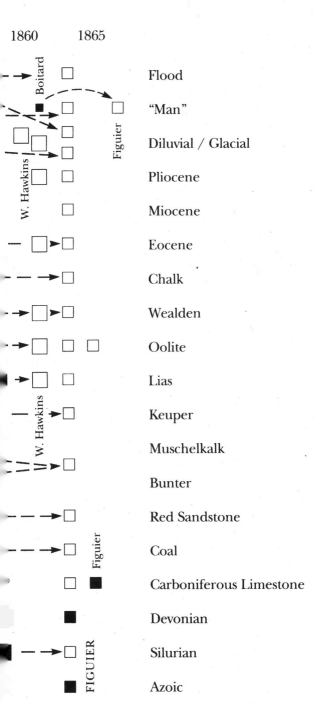

1860   1865

Boitard

Figuier

W. Hawkins

W. Hawkins

Figuier

FIGUIER

Flood

"Man"

Diluvial / Glacial

Pliocene

Miocene

Eocene

Chalk

Wealden

Oolite

Lias

Keuper

Muschelkalk

Bunter

Red Sandstone

Coal

Carboniferous Limestone

Devonian

Silurian

Azoic

chapters (fig. 106). As in any piece of history worth the name, a close attention to chronology—even if fashionably dismissed as "mere" *histoire événementielle*, a "mere" chronicling of events—brings its own rewards. Figure 106 plots most of the scenes that have been described and reproduced in this book onto a two-dimensional field of time scales. On one dimension is the scale of years through the middle decades of the nineteenth century. On the other dimension is the unquantified relative time scale of earth history, which is perhaps the greatest legacy that modern science has inherited from the geologists of that period (the time scale is shown with its nineteenth century names, but most of them will be recognizable to modern geologists).

Though it should cause no surprise, this diagram highlights at once the cumulative character of the story. Unger's great sequence, arrayed appropriately at the center of the human-historical time scale, is preceded by a record of much more fragmentary pictorial experiments. De la Beche's highly innovative scene of Liassic life, roughly at the center of the geohistorical time scale, stands at the start of the most prominent earlier lineage. It is joined successively by scenes based on Cuvier's earlier reconstruction of the bodies of Eocene mammals, on the dinotherium from somewhat younger Tertiary strata, on the "Diluvial" mammals, and on the fossil flora of the Coal strata: a list that is quite random in relation to the geohistorical periods represented. Later, Unger's primarily botanical sequence is amplified by Hawkins's full-sized models of Secondary reptiles and by Unger's own supplementary scenes from the oldest strata with *any* fossils. Finally—as far as this book goes—all these earlier resources are consolidated into Figuier's fuller series of scenes.

This cumulative historical record justifies the repeated mention of "firsts" in the narrative chapters of this book.

Simply as historical description, the first examples of scenes of particular kinds have been pointed out with some emphasis, because they constituted important enlargements of the repertoire of pictorial resources that was available thereafter to others. In no way do they imply a teleological direction or inevitable path of "progress" toward some kind of fulfillment in Figuier's work or any other—the justification for closing the narrative of this book with Figuier will be considered at the end of this chapter—and certainly not toward an unambiguously truthful or "realistic" way of depicting the deep past. On the contrary, the history of these scenes is permeated at every level with the contingencies of their human construction: in all its details and in most of its larger features too, the story might have developed otherwise.[4]

One element for which some degree of inevitability might be claimed, however, is in the relation between the development of scenes from deep time and the parallel developments in stratigraphical geology and paleontology on which the scenes were necessarily based. Only the most extreme skeptic would doubt that the record of the strata is a generally reliable—if not unproblematic—record of a sequence of real events in the history of the earth. But even here the relation between the scenes themselves and their underlying sciences is far from straightforward.

Most of the stratigraphical "succession," or series of "formations" of strata, had been established (with the degree of detail relevant to this story) long *before* the corresponding scenes were first produced. Of course, geologists in the early nineteenth century were well aware in principle that their series of stratigraphical formations was a record of a sequence of periods; but in practice they treated it more often as a *structural* stack of three-dimensional rock masses than as a *temporal* sequence of events in earth history. As late as the 1820s, the sheer novelty of any fully *historical* reconstruction of the deep past is vividly expressed, for example, in Conybeare's cartoon and doggerel celebrating Buckland's verbal reconstruction of an "antediluvian" hyena den (fig. 17; text 10). Likewise, by the time Unger extended his original sequence of scenes to cover still earlier periods (figs. 67, 68), the corresponding Silurian and Devonian "systems" of strata had been defined (mainly by Murchison) and recognized internationally for almost two decades.[5] The description of the relevant strata and their fossils was obviously a necessary condition for the construction of any corresponding "scene"; but equally clearly, it was not a sufficient condition. Something more than outcrops of well-hammered rocks and trays of well-handled fossil specimens was needed before any pictorial scene of the world at the time of their formation became—in the literal sense—conceivable.

This is where our own familiarity with scenes from deep time can blind us to the novelty and problematic character of their conception. The first such scenes clearly borrowed from pictorial precedents in both natural history and biblical (and more generally, historical) illustration. But however helpful those artistic traditions may have been, as resources for the new kind of picture, they failed at one crucial point. Scenes of present-day animals and plants in their natural habitats were scenes that naturalists had witnessed, or at least that they believed they *could* witness if they had the opportunity to travel to an appropriately exotic part of the world.[6] So the conventional human viewpoint of such scenes—drawing in turn on pictorial precedents in landscape and topographical art—reflected a tacit claim that if one could observe them for oneself, the animals and plants really would look the way the artist had depicted them. Likewise, scenes from sacred or secular history were based on texts that the artists believed were trustworthy accounts of events that the

original authors (or at least *their* sources) had actually witnessed. Again, the conventional human viewpoint reflected a tacit claim that the events really had looked to
the original participants something like what the artist
depicted. In both cases, the construction of a pictorial
scene with a human viewpoint was epistemically justified
by its relation to a scene in the real world that had been,
or could be, witnessed by human beings.

Scenes from deep time, by contrast, had not been and
could not be witnessed by any human beings at all. By
definition they claimed to represent a human viewpoint
in a world that was totally nonhuman because it was
prehuman. Here, significantly, the only precedent that
might have been helpful was one with which many nineteenth-century "men of science" were reluctant to be
associated. Traditional biblical illustrations had always
included scenes from the very beginning of time, before
any human beings were present to record the events
depicted. Scenes such as those in Scheuchzer's book depicted the "days" of Creation (figs. 1–6) from exactly the
same human viewpoint as later scenes, such as those of
the Deluge (fig. 7), that were taken to have had real
human witnesses (even if those outside the Ark, whose
viewpoint was represented, had not survived to tell their
tale!).

However, that pre-Adamic human viewpoint was not
a divine viewpoint, unless it was that of the anthropomorphic Yahweh "walking in the garden in the cool of
the day," or, proleptically, that of the *logos* that "was made
flesh and dwelt among us." If a true Creator's viewpoint
was needed, there was always the established convention of a view from the heavens (in modern terms, from
outer space), as, for example, in Burnet's well-known
frontispiece of the past, present, and future states of the
earth, with cherubim as onlookers.[7] The only effective
precedents, then, for prehuman scenes with a human

viewpoint were those of biblical illustrations of the Creation story. That very fact may help explain the apparent reluctance of "men of science" to construct analogous scenes, even if they were based on the new evidence of geological science rather than on biblical texts. Conversely, the same fact may explain why that reluctance was first overcome, and science-based scenes first constructed, in the one major European country where—as contemporary observers often noted—the practice of religion was not regarded as antithetical to the practice of science. Buckland and Conybeare, for example, were ordained ministers of the Church of England; and even if De la Beche was anticlerical in attitude, he could not but be affected by the religious dimension of the culture of his social class.

However, even if a specific cultural context facilitated the initial conception of the possibility of constructing scenes from prehuman time, the paradox of their human viewpoint remained. That viewpoint had to be justified and explained in secular terms that would be acceptable throughout the world of scientific culture. This required the invention of some mode of imaginary time travel, analogous to the real spatial travel of a naturalist on a voyage or expedition. Again, our own familiarity with this idea—thanks to a fictional tradition extending from before H. G. Wells to Dr. Who and beyond—can keep us from seeing the startling novelty that its first scientific embodiment entailed. That sense of novelty, as felt at the time, is again well expressed in Conybeare's visual and verbal jokes about Buckland's hyena den (fig. 17; text 10). Buckland was portrayed as *penetrating* in time from the human world of the 1820s into the prehuman world of antediluvian Yorkshire. The physically narrow entrance into the cave was perceived as an epistemically narrow opening into the deep past: as Conybeare expressed it, with pardonable poetic exaggeration, "What was done ere the birth-day of time, thro' one other such hole I could spy."

What was spied was of course seen from an implicitly human viewpoint: in Conybeare's cartoon, it is as if Conybeare himself has preceded Buckland into the cave and is there among the hyenas to record the moment of his colleague's arrival. The unseen artist's eye imports a tacit human presence, like the unseen Ansel Adams behind a photograph of a wilderness apparently empty of humanity. Likewise, the great majority of later scenes are in the form of *landscapes:* prehuman scenes depicted as if a landscape painter such as Kuwasseg had set up his easel in a Coal forest or among the dinosaurs (figs. 44, 51), instead of in his native Austrian countryside.

This pictorial convention explains the obvious links with the traditions of ordinary landscape art. But even here the adaptation was not as straightforward as it was made to appear, because most scenes from deep time— Conybeare's cartoon is here an exception—claimed to represent an idealized landscape from a specific period, not any particular locality. However, there was one well-established and still lively tradition within landscape art, which artists such as Kuwasseg and Riou could use as a resource for their explicitly "ideal landscapes." Imaginary landscapes had long formed the conventional backgrounds for scenes from classical mythology or biblical history; even when detached from such subjects, landscapes *not* based on a specific locality continued to be more highly prized in the artistic world than those with a "merely" topographical purpose.[8] The depiction of an "ideal" landscape for, say, the period of the Chalk thus had well-established artistic precedents, even if it was a daunting task scientifically.

However, the artistic traditions with which scenes from deep time had their closest links were those in which the term "scene" has its full theatrical meaning: not "just" a landscape, but a landscape with figures, or a scene with a specific action in progress. These were pictures in which the "figural" content was outweighed by the "discursive."[9] Characteristically, such scenes were not considered to be self-explanatory, even if the plot or outline of the action depicted was widely known. John Martin, for example, published a pamphlet to explain his picture of the Deluge, even though the story in Genesis was universally familiar to his public (fig. 10; text 7). The figural elements that locked the picture into the familiar story had to be located and identified (even the Ark did, but only because it was depicted so small and distant!). What Martin presented to his public was a visual representation linked to an indispensable textual explanation. That kind of linkage was adopted unchanged by most of those who later constructed scenes from times even deeper than the Deluge; it has deliberately been reproduced in this book, by the close conjunction of figures and texts. The publics that learned to "read" the new pictorial genre did so, as they were used to doing, in conjunction with textual explanations. Neither was "parasitic" on the other; the appropriate biological metaphor would be that of symbiosis.

The adopted convention of a tacit human viewpoint, then, enabled scenes from deep time to be understood quite readily in the light of a familiar conjunction of visual and verbal practice. But that viewpoint did entail one serious problem. Paleontologists were agreed that the great majority of fossils were the remains of organisms that had lived in the sea, and wholly underwater at that. Any adequate visual representation of the deep past therefore had to include some way to depict that underwater world. Even the present-day relatives of these organisms could not easily be seen in their natural habitats, before the days of snorkeling and scuba diving. So the problems of picturing a prehuman world were compounded by those of depicting a world that was also non-human because it was *subaqueous*.

For the most part, however, that problem was simply evaded. Most scenes from deep time, within the period covered in this book, portrayed ordinary marine organisms as having been washed up on a shore, in the foreground of a landscape seen unproblematically from a human viewpoint (e.g., figs. 40, 52). In this respect, they simply continued the established pictorial convention represented, for example, by Scheuchzer's and Parkinson's earlier scenes of the biblical Deluge (figs. 8, 9). In effect, the aquatic world from which most fossils were derived was depicted only from the outside, from the *subaerial* world to which a time-traveling human observer could more plausibly have had access.

However, the very first true scene from deep time offered a radical alternative. De la Beche's *Duria antiquior* (fig. 19) was constructed from a pictorial viewpoint right at sea level, offering a dual view—like the bifocal vision of some fish—of both the subaerial and the subaqueous worlds. This imaginative formula, which is too early to be derived from the experience of peering through the front panel of an aquarium, may simply come from the habit of messing about in boats and rock pools on the Dorset coast, or perhaps from contemporary diving practices for salvaging underwater wrecks. In any case, what is significant is that it was not fully adopted in *any* later scene reproduced here, apart from one anomalous exception in Figuier's book (fig. 83); and even that was an afterthought, substituted in later editions for a more conventional earlier picture (fig. 82). This suggests how difficult it may have been for the public to which all these scenes were directed, and perhaps for most of the geologists too, to *imagine* a viewpoint that was not only prehuman but also subaqueous—at least until mid-century, when the famous aquarium craze made the underwater world generally accessible for the first time.

The fundamental problem of representing the pre-human world of deep time, aggravated by that of representing the subaqueous world of *any* age, must surely account in part for the apparent reluctance of early nineteenth-century geologists to incorporate these imaginative exercises into their ordinary scientific publications. Their reservations about the speculative character of pictorial scenes were compounded by the social pressures on the self-consciously new science of geology, which needed to prove its scientific credentials and political soundness by publicly emphasizing and exemplifying its soberly factual basis. The description and ordering of strata and formations and the identification and classification of animal and plant fossils—these were regarded as activities worthy of a science. An imaginative reconstruction of what the world would have looked like to a human observer, long before there were in fact any human beings there to do the observing—that seemed to smack too much of the unbridled fantasies of an earlier age, which had given the study of the earth a bad name.

The earliest scenes from deep time were therefore introduced into scientific discourse in an unmistakably marginal manner. Cuvier's multivolume work on fossil vertebrates, which set the pace for paleontology for the rest of the century, included only one set of pictorial reconstructions of body profiles (fig. 16). Even those drawings were not his, but his assistant's; they were strikingly inferior to his own unpublished reconstructions of the same mammals (fig. 15), and less informative than his verbal reconstructions of their likely appearance and habits (text 8); and they were never synthesized into any kind of landscape scene. Buckland likewise produced a vivid verbal description of the immediately "antediluvian" world (text 9), but it received visual expression only in Conybeare's jokey cartoon (fig. 17). In his later

Bridgewater Treatise—otherwise a masterpiece of *haute vulgarisation*—Buckland included only one tiny scene (fig. 29), and that was borrowed from Goldfuss. Goldfuss's first attempt at a scene was literally squeezed into the margins of a conventional illustration of fossil bones (fig. 21); he eventually included two major scenes in his great monograph on fossil invertebrates (figs. 22, 40), but neither had any real relevance to the rest of that work. De la Beche created his scene of Liassic Dorset (fig. 19) as a privately distributed and semi-humorous performance to benefit an impoverished professional fossil collector; but in his serious published work, he rejected its impressive format and highly innovative design in favor of far more timid and unrealistic little illustrations (figs. 23–25). When Kaup discovered a striking new specimen that made possible a full reconstruction of the largest land animal then known, he commissioned a fine scene to show it in its likely habitat (fig. 30); but he placed it as a decorative vignette on the cover of his monograph, not in the body of the work itself; and he paired it with a scene of the excavation, featuring as a joke the ghost of the extinct animal itself (fig. 31). Years later, in the first serious scientific work devoted exclusively to a sequence of pictorial scenes, Unger still felt it necessary to apologize for the speculative character of the undertaking (text 31).

The reluctance of all these authors to give more than a marginal place to scenes from deep time may have been compounded by the sense that they could not fully control the product. Although they lived in a culture in which some artistic competence was one of the marks of education, most of them lacked the skills to translate their mental and verbal images of the deep past into a visual form adequate for publication. It may be no coincidence that the finest early scene was produced by one geologist (De la Beche) who did possess substantial artistic skills, and that his only rival in that respect (Cuvier) was an outstanding zoological artist. Apart from such exceptions, would-be composers of scenes from deep time were obliged to commission someone who did have the necessary artistic skills. By doing so, however, they inevitably relinquished the complete control over their material that they enjoyed when expressing their ideas or results in verbal form.

Most of the earlier scenes from deep time were in fact drawn by professional artists—usually landscape painters or book illustrators—who generally received no more acknowledgment than was customary for such work. Their names were inscribed modestly at the foot of the print, along with that of the engraver or lithographer, and that was that. Just how the collaboration worked is difficult to recover, beyond clues such as the note that Thomas Hawkins, for example, "designed" the scene that Templeton drew for him (fig. 28)—probably a rough sketch, but possibly no more than a written or even spoken suggestion.

In such cases, the artist was nearly as invisible as other technicians have generally been in science, past and present.[10] Only in a few cases did the geologist deign to acknowledge in print the debt he owed to the artist who had realized his vision of the deep past. Among such exceptions are Richardson's appreciation of Nibbs, and Trimmer's of Whichelo (texts 27, 28). Mantell's and Thomas Hawkins's recognition of John Martin's work (texts 25, 26) falls in a special category, for here it would have been self-defeating—and bad for sales—*not* to highlight the achievement of co-opting such a widely known and distinguished artist. Even if Martin received a fee for his work, as doubtless he did, it was really he who was passing credit to his patrons, rather than the other way round.

It was only with Unger's collaboration with the landscape painter Kuwasseg that anything like a partnership of equals was established in such projects. Here too, the artist presumably received a fee for his work, but Unger's praise for his contribution and description of their collaboration sound heartfelt (text 31). Again there is little evidence about the form that the collaboration took, beyond the hints of a protracted dialogue between the two partners, in the course of which the artist's efforts came to embody satisfactorily the botanist's mental and verbal imagery. That kind of partnership of equals was reproduced later when Figuier, with Unger's example before him, engaged Riou to draw scenes for him, and gave his work similar prominence. Such collaborations between scientists and artists—and, often, many other technicians—continue of course to be vital to the production of modern scenes from deep time, particularly in the technically complex three-dimensional dioramas that decorate our modern museums.

Most of the earlier scenes from deep time were directed toward their authors' peers. They were produced for other "professional scientists"—though that modern term is seriously anachronistic until after the end of the period covered in this book—or at least for serious "amateur" fossil collectors and other gentlemanly "men of science" (and a few women). But that original audience for these scenes was quickly supplemented, and soon superseded, by much broader and more popular ranges of readers, or rather, viewers. Some leading "men of science" continued to contribute to the genre, if not directly then at least as advisers to more popular authors or artists: Owen's role as adviser to Waterhouse Hawkins on the Crystal Palace exhibit is an outstanding example (fig. 61). Doubtless, many others continued to study, evaluate, and criticize the scenes published in popular works. But the locus of the genre shifted unmistakably away from the specialist literature into more popular channels. There it has remained ever since: in expert discourse, abstract or schematic representations have come to be regarded as more useful, for conveying scientists' interpretations of the deep past to each other, than the "realistic" scenes that appeal so widely to the general public.[11]

The "popularization of science" was formerly treated as a wholly one-way process, *de haut en bas,* by which scientific pundits translated—or allowed others to translate—their esoteric findings into more accessible language, with inevitable loss or distortion of content on the way. More recently, however, the process has come to be seen as being initiated as much from the "popular" end as from the "scientific," from the demand end as much as from the supply. Even if it is initiated from the supply end, the "knowledge" may be profoundly—and often intentionally—transformed on the way, by being used in the service of social objectives that have little to do with the supposedly disinterested character of science.[12]

Even more illuminating than studies of popular science, however, is the art-historical work that has directed attention to the symbiosis between painter and patron, not only in economic terms but also in terms of shared visual conventions and reciprocal visual responses.[13] Up to this point, the development of the genre of scenes from deep time has been described in terms of pictorial precedents that enlarged the repertoire of visual resources that were available to "men of science" and their artists. But it should go without saying that the use of those resources was at the same time part of a process by which those who viewed scenes from deep time came to share a common visual language with those who had constructed them. To make their scenes intelligible, the artists had to use conventions that were already familiar to their viewers. The adaptation of pictorial models from natural history or biblical illustration was thus not only a way of constructing adequate representations of unfamiliar subjects; it also ensured that scenes from deep time would be intelligible to those who studied them. From this perspective, the widening of the circle of viewers, from "men of science" to the general public, entailed no sharp break of practice, but simply a continuation of the same process of developing a set of visual conventions that would make the scenes intelligible to an ever-broadening range of viewers.

After De la Beche's innovative portrayal of Liassic Dorset (fig. 19) had been widely distributed, albeit in semi-private form, and after Goldfuss's amplified version (fig. 22) had been published in Germany, its potential appeal to a much wider public was quickly appreciated. Modified versions soon appeared in both English and French publications distributed by subscription (figs. 26, 27); in the former case, with a truly *mass* public. In both cases, however, the authors (Phillips and Boblaye respectively) were not journalists or publishers—or, in modern terms, "science writers"—but geologists with substantial scientific reputations in their own right. It was they who realized the popular potential of illustrations that their publishers, on their own, might never even have known about. On the other hand, it was of course the educational and commercial imperatives of their respective outlets that led Phillips and Boblaye to think of adapting the pictures for a wider public. If only for commercial reasons or to maintain a uniform style, their scenes had to conform to the format of other illustrations in the same publications (as black-and-white wood engravings and colored steel engravings respectively). But the association with more prosaic illustrations also served to impute an unproblematic "ordinariness" to scenes from deep time. Thus although the initiative came from those who were later termed "scientists," the first scenes from deep time were quickly assimilated into an already flourishing culture of popular science, and specifically into the popular fashion for natural history.[14]

Soon afterward, the genre was appropriated by an important specific branch of popular science, namely, science for the young. The two modest but imaginative scenes by Peter Parley (figs. 32, 33), like those just discussed, suggested the ordinariness of deep time by assimilating the scenes to the same style as other less problematic pictures. From that point on, children became a major audience for scenes from deep time, and not only in works such as Parley's and Figuier's that were explicitly directed to them. Certainly cartoons evoked by the Crystal Palace exhibit and by Figuier's book (figs. 64, 105) suggest that the youthful appeal of scenes from deep time—even if it was the fascination of the horrible!—was not only in the minds of their elders and betters. That appeal has of course continued to characterize the genre ever since.

However, perhaps the most important medium by which such scenes were appropriated into the culture of popular science was the ordinary introductory book on geology. Of these there was an astonishing number in the middle decades of the nineteenth century: the science had a breadth of appeal then that it had never had before, and arguably has never had since (unless the recent dinosaur craze is counted as geology). At least a single scene from deep time, usually centered on the Jurassic reptiles, became almost a cliché in these popular books. Among the small sample that has been reproduced here, those by Richardson, Zimmermann, and Brewer (figs. 38, 57, 58) exemplify the general style. Placed usually as a frontispiece—the nineteenth-century equivalent of modern dustjackets—these scenes served to whet the appetite of potential purchasers, giving them a tempting glimpse of the "wonders of geology," which was often hardly fulfilled by the dry contents of the rest of the book. In terms of style and content, they were generally, and unsurprisingly, derivative; but collectively they evidently generated a vivid sense of the "wonder" of deep time in the literate mass public that was emerging at just this period. It is no coincidence that that same public was also becoming aware of the "romance" of *human* history, not least of the "otherness" of the Middle Ages.[15]

That broad public for scenes from deep time transcended national and linguistic frontiers, and also, at least to some extent, the more solid nineteenth-century barriers of social class. The genre showed little modification as it crossed cultural boundaries. Economic considerations alone were sufficient incentive to reuse pictures in editions in different languages or in publications aimed at different audiences, just as they are today when television programs are moved across national or linguistic boundaries with no more than dubbing or subtitling to adapt them. Translations of standard texts such as De la Beche's and Mantell's carried their respective scenes from English- to French- and German-speaking audiences. Unger's bilingual German and French texts made his great sequence of scenes accessible immediately—in principle if not in practice—to most of the scientific world, while Highley's apparently pirated editions soon brought it to the insular anglophone world as well. Hawkins's full-sized models of extinct reptiles were seen at first hand by the multinational public that flocked to the Crystal Palace, and indirectly through innumerable newspaper reports and illustrations of the exhibits there. Riou's engravings in Figuier's book were simply used again and again, when the French text was translated into other languages; and they were adopted by Fraas—doubtless by commercial arrangement—for his similar book for the German-language market.

In none of these works was it thought necessary to make any modifications to the scenes, or significant ones to the texts that explained them, in order to adapt them to different national cultures. Of course, that can and probably should be taken simply to reflect the cultural homogeneity, throughout the Western world, of the middle-class readers for whom most of these works were primarily designed. But Phillips's scene for the *Penny Magazine* (fig. 26) suggests that no adaptation was deemed necessary even for a mass audience that extended down the social scale at least as far as the literate artisan class. Likewise, Waterhouse Hawkins's reptiles at the Crystal Palace were seen by a far wider spectrum of Victorian social classes than their living counterparts at the London zoo.

Even the single scenes that decorated so many popular and introductory books on geology served, then, to construct for their readers some sense of the vanished reality of the deep past. As already suggested, some of the earliest scenes tended to impute an unproblematic "ordinariness" to deep time, simply because they were presented in the same style and format as illustrations of the present world. But the accompanying texts reveal that many of the "men of science" who drew or commissioned the scenes regarded the denizens of deep time as profoundly alien, or at least that was the impression they chose to present to their readers. The Jurassic reptiles seemed to Boblaye like the products of "a diseased imagination" (text 17); Goldfuss likened the flying pterodactyle to "the unbounded fantasy of a Chinese artist" (text 14), while Buckland compared it to Milton's fictional "fiend" (text 19). Seen in this light, the bizarre effusions of Thomas Hawkins, in his description of an "eltrich-world uninhabitate, sunless and moonless" (text 18), are rather less removed from mainstream science than might at first appear.

With this background of Romantic or "Gothick" imaginings, Mantell's strategic recruitment of John Martin to create a visual scene from deep time (text 24) was likewise less of a innovation than Martin's style might suggest.

241

Mantell—again, a "man of science" in his own right—perceived correctly that a frontispiece by Martin would not only help the sales of his popular book, but might also express in adequate visual terms the widespread sense of the otherness of the vanished world revealed by geology. That world was perceived as alien because its animal inhabitants were entirely extinct; more, because some of them were not only "monsters" in the technical sense (combining characters now defining separate classes) but also seemed so in the everyday sense; but most of all, because their world had been wholly lacking in the human presence. The sinister darkness of Martin's designs for Mantell and Hawkins (figs. 35, 36), and the nightmarish dragons into which the fossil reptiles were transmuted, displayed the deep past as a world profoundly different from the present, as different as the cosmic chaos in Martin's earlier picture of the Deluge (fig. 10). By association, the "country of the iguanodon" was equated with the mythical landscapes of countless earlier depictions of "Saint George and the Dragon."

In that full-blooded form, Martin's impact on the genre of scenes from deep time was short-lived. Richardson, for example, reverted in later editions of his introductory book to the kind of scene he had used before his former associate recruited Martin (figs. 34, 38)—and probably not only because Mantell's scene by Martin was much more striking than his own (fig. 37). Quite generally, there was an inevitable tension between the melodramatic attractions of the Martinesque style and the more prosaic claims of geology to be a science founded on sober induction from concrete facts. Martin played fast and loose with the anatomy of his reptile-dragons, in a way that authors other than Mantell and Hawkins probably felt was counterproductive. Most later scenes therefore reverted to a style that was far less lurid, and far closer to contemporary landscape art of a pastoral character.

Nonetheless, some trace of Martin's style can still be detected in many later designs, such as the more stormy scenes in Kuwasseg's sequence for Unger, or his scene of the great Jurassic reptiles at each others' throats (figs. 48, 51). In fact, it seems to persist to this day in the public—and particularly the childish—perception of the deep past as a realm of nightmarish horrors and frightful "prehistoric monsters." The cartoonists' responses to Waterhouse Hawkins's full-sized models and to the images in Figuier's book (figs. 64, 65, 105) were probably not without some basis in their children's experience (and perhaps in their own too!). The persistence of this emotional response to the worlds of deep time must surely account in part for the continuing popularity of dinosaurs, indeed, the increasingly cultlike fascination with them. Whether this had, and still has, tacit functions—cultural or depth-psychological—and what they may be are questions that would take this book beyond its proper scope, and this author out of his depth.

Another enduring legacy of the transient Martinesque episode was the way it reinforced, in striking visual terms, the sense of a *single* alien world of the deep past. The geologists had worked hard to distinguish many successive "systems" of formations, each with a distinctive set of animal and plant fossils. But among the general public, the older concept of a single undifferentiated world buried in the strata died hard. The well-worn singular phrases—"the ancient world," *Die Urwelt, l'ancien monde*—continued their useful life, even in the titles of works by the geologists themselves. Unger's work, the first to be devoted wholly to a *sequence* of scenes, was entitled *The Primitive World*, even though that singularity was immediately qualified by a reference to its "different periods of

formation." Figuier's best-selling book likewise advertised the sequence of diverse scenes it contained, yet it
was entitled *The Earth before the Deluge*. Fraas's similar
book was given a cover design in which the creatures of
all the different periods were assembled in a single scene
(fig. 104).

Since the public persisted in regarding the deep past as
a single undifferentiated world, the marketing strategy
behind such decisions was astute. Despite Waterhouse
Hawkins's design to instruct the public otherwise (fig. 63),
his spectacular exhibit at the Crystal Palace was perceived
by the cartoonists—and doubtless by those who were
amused by their work—simply as a resurrection or nightmarish vision of *the* "antediluvian world" (figs. 64, 65). In
the context of the imperial celebration that characterized
the Crystal Palace exhibits as a whole, the scientific reconstruction of that alien realm was as much an expression
of successful *conquest* as the caging of fierce exotic beasts
in the zoo in Regent's Park.[16]

That public conception of a single vanished world of
"prehistoric monsters"—which persists widely to the
present day—is closely analogous to the popular conception of the human past as undifferentiated "olden days,"
to which "heritage" exhibits and historical "theme parks"
now implicitly claim to give direct and unmediated
access.[17] Just as the latter sharply separate the lived-in
present from the "history" embodied in all earlier human
generations, so the deep past was regarded as sharply distinct from the present world, primarily because it was
nonhuman. That geologists should claim to be able to
portray the farthest reaches of earth history as if they had
personally been there is, of course, what made it so striking to the general public. So it is not surprising that over
some of the earlier scenes there hangs the aura of either a
fairy tale, as in Conybeare's cartoon of Buckland's hyena
den (fig. 17), or a nightmare, as in Martin's scenes for

Mantell and Thomas Hawkins (figs. 35, 36). The almost incredible and implicitly nonnatural character of such time travel became a recurrent theme. In Boitard's popular book, for example, it involves explicitly the magic of Le Sage's "lame demon," who conducts the nineteenth-century author (seated on a meteorite!) back into the time of the plesiosaur (fig. 76; text 58). Only in twentieth-century science fiction has the time travel implicit in any scene from deep time been transmuted yet again, as befits a pervasively technological age, into the gadgetry of time *machines*.

Time travel in turn implies the possibility of a journey with stations on the way. But the earlier scenes from deep time, with few exceptions, remained isolated from each other. They were single scenes based on various parts of the geological succession. They were not deliberately selected as representative of the successive periods of earth history; rather their choice was enforced by the contingencies of particular fossil discoveries on which scenes could be based. Buckland's proposal for a sequence of scenes (text 15) was not followed up; and other innovative experiments in portraying the whole temporal dimension in scenic form, such as the pile of scenes that Whichelo drew for Trimmer (fig. 39) and the ingenious continuous panorama that Emslie drew for Reynolds (fig. 42), seem to have had little impact. Only with Unger's commissioning of Kuwasseg was there a full-scale attempt to portray, in true geohistorical order, *every* period for which there was adequate evidence on which to base a scene (figs. 43–56).

The format of a sequence of discrete scenes was suggested, then, by Buckland, tried out modestly by Trimmer, but only exemplified definitively in Kuwasseg's magnificent series for Unger. Thereafter, and particularly as a result of its adoption by Figuier (explicitly from Unger), it became the canonical form for the genre for the rest of the nineteenth century and the whole of the twentieth. What remains at present no more than a plausible inference is the claim that its close similarity to the long-established genre of biblical illustration is no coincidence. Unger himself was the product of a deeply Catholic culture, and the text for his culminating scene (text 46) contains strong biblical imagery. More significant, however, his sequence of scenes is structured around a concept of linear historical *development,* in which the successive periods of the deep past find their culmination, and perhaps their fulfillment, in the human world of the present.

That teleological sense of the purposive directionality of earth history and the history of life is of course closely analogous to the traditional Judaeo-Christian interpretation of human history, including the brief prelude that was traditionally prefixed to the human story (in the first chapter of Genesis). From this perspective, the close pictorial analogy between Kuwasseg's sequence of scenes for Unger (figs. 43–56) and Pfeffel's sequence of scenes for Scheuchzer more than a century earlier (figs. 1–6) is no surprise. Of course, Unger did not share Scheuchzer's taken-for-granted adoption of the traditional short time scale of cosmic history. Like all his geological contemporaries, Unger was well aware that the sequence of strata must represent literally unimaginable periods of time, dwarfing beyond measure the totality of human history. But he—again like most of his contemporaries among the geologists—retained nonetheless the sense of purposive linear development, stretching all the way from the earliest known life to the human world.

The portrayal of a single scene, and even of a sequence of scenes, did not strictly compel assent to the geologists' sense of the vast scale of deep time. In practice, this was one of the most important points on which scientific opinion had diverged from popular beliefs. Though it is difficult to document satisfactorily, popular opinion seems to have long continued to take the traditional short time scale for granted: much of the general public—particularly in the anglophone world and in the earlier part of the century—still learned its cosmic history in church, or from the family Bible. Probably this reinforced the tendency to regard the whole of the past disclosed by geology as a single "primitive world." Specifically, it helps explain the persistence of the imagery of the Flood as the symbolic boundary between the deep past and the human world, as in the continuing use of the traditional term "antediluvian" and the continuing appeal of titles such as Figuier's *Earth before the Deluge*. Of course, the antediluvian world had not been strictly prehuman, but the shadowy character of *all* the events in the early chapters of Genesis—from Creation through Adam to Noah—made that traditional narrative a plausible match for the equally shadowy "wonders" displayed by the geologists.

Even before a full sequence was widely available, however, scenes from deep time had become a powerful form of visual rhetoric, with which the wider public could be persuaded of the reality of the geologists' unimaginably long time scale. Mantell emphasized the "myriads of years" and "countless ages" that had elapsed since the time of the iguanodon (text 25); Zimmermann, alluding to his similar scene, likened the whole of human history to "a mere zero" in comparison to the age of the earth (text 47). By the time Waterhouse Hawkins's full-sized models were displayed to the public at the Crystal Palace, this sense of the time scale had evidently been absorbed quite generally: one journalist who described the exhibit referred to the oldest known human history as being "as of yesterday" by comparison with the world of the extinct reptiles (text 51).

This growing public sense of the vast time scale portrayed by the geologists' scenes was surely facilitated by the fact that it entailed no necessary break with the traditional religious meanings of the Genesis narrative. The public could simply adopt the newer understanding of biblical language that had been made available by the slow spread of biblical criticism. Like Thomas Hawkins, commenting on his scene of "the pre-Adamite—the just emerged from chaos—planet," an appreciation of the humanly immeasurable periods involved could easily be combined with allusions to the traditional "days" of Creation, understood now as vast periods of prehuman time (text 18).

Thus although scenes from deep time did not demand an acceptance of the geologists' time scale, they certainly provided a powerful rhetorical means to persuade the public of its veracity. Likewise, the format of a sequence of scenes, illustrating the temporal development of the living world, did not demand that the history of life be given an *evolutionary* interpretation or imply that it had been smoothly continuous. For a sequence of scenes could be interpreted impartially in either of two ways. It could indeed be regarded as a series of sample moments plucked from a smoothly continuous panorama of change, like stills cut from a movie or frozen on video (in the nineteenth century, the proliferation of toys that created the illusion of a moving scene from a sequence of stills made this as familiar an idea as it is today). But alternatively, it could be seen as a series of discrete "scenes," in a sense close to the theatrical use of the word, separated from each other by abrupt (and even perhaps nonnatural) changes of scene.

On the other hand, the format of a sequence of scenes certainly allowed for the possibility of an evolutionary interpretation of the history of life. Probably it encouraged it, by vividly displaying the scale of the changes for which evolutionary theories claimed to provide an explanation. Unger, for example, seems have assumed some kind of natural process of organic change, and Figuier's text is full of evolutionary turns of phrase. In any case, however, the genre later became a powerful and explicit form of visual rhetoric in the service of evolutionary theories; and it has continued to be used for that purpose throughout the twentieth century, in popular science books, television programs, and museum displays.

The scene that first claimed to set the earliest *human* life into the visual context of still deeper time was, unsurprisingly, the final scene in Kuwasseg's sequence for Unger (fig. 56). For just as a scene of Adam in Eden had formed the culmination of Pfeffel's sequence for Scheuchzer's Creation narrative (fig. 6), so the arrival of a human presence defined for Unger the culmination of the similar but immeasurably longer story revealed by geology. Although Kuwasseg's design was unmistakably Edenic in tone, and certainly not modern, Unger entitled it "The Present World," because it portrayed a human presence; in this respect it stood in contrast to all the scenes of the prehuman "primitive world" that preceded it. At the same time, however, its stylistic similarity to all those earlier scenes made it appear as the plausible culmination of the longer history. In fact, that tension between human and prehuman is reflected pictorially in the way the human family stands out sharply against its nonhuman background.

Perhaps deliberately, that Edenic, or at least Arcadian, vision ignored the growing suspicion and accumulating evidence that the history of humanity had in fact been much more fully embedded in that of the natural world. Specifically, by Unger's time it was being claimed that the first human beings had been the contemporaries of the last of the spectacular extinct animals that paleontology had recovered. By the time Boitard's posthumous work prompted Figuier to try his hand at the same kind of book, a debate on that issue was in full swing, and the two authors expressed their opposing opinions in scenes of strikingly contrasted content. Figuier opposed the newer view by getting Riou to draw an Arcadian scene much like Kuwasseg's (fig. 101). This was implicitly designed to counter Boitard's nightmarish vision of human ancestors even more bestial in their habits than the apes they physically resembled (fig. 77). In his later editions, however, Figuier was forced by a hardening consensus of the experts to make some concessions; and Riou's later scene, like Boitard's, showed the first human beings as cave dwellers clothed in skins and armed with stone axes, confronting a wild and hostile animal nature (fig. 102, compare fig. 78).

Even here, however, Riou's human beings remained as white, European, and civilized in demeanor as the family in Kuwasseg's picture and his own earlier design. Morally, if not technologically, they were still familiar and reassuring to Figuier's vast middle-class Western public: they were *themselves* dressed up as primitives, though hardly more convincing than Marie Antoinette as a milkmaid. At this point, the genre of scenes from deep time was drawn into the service of social and even racial goals as blatant as those that had long characterized the portrayal of "savages." But that imputation, if it is to be fair, must be symmetrical: social interests of the same kind, albeit opposite in intention, were also and equally served by Boitard's deliberately unflattering picture of our remote ancestors.

What these scenes show is that the human presence could not be assimilated visually into the story of deep time without being co-opted to convey implicit messages about the relation between humanity and the natural world, between nature and human nature. The genre in itself remained neutral, just as it was in relation to the time scale it reflected and to evolutionary explanations of the changes it exhibited. But in each case, it proved to be a powerful visual rhetoric for those with the power to impose their views on the wider public. That ultimately *political* use of scenes from deep time is first clearly seen in Kuwasseg's, Boitard's, and Riou's renderings of the earliest human beings; it continues unabated, perhaps indeed with increasing force, in our present-day museums and television programs.[18]

In conclusion, the limitation of this book to the *first* scenes from deep time needs further explanation. One practical reason, of course, is that to trace the development of the genre after the 1860s and up to the present day would have entailed a doubling, if not a tripling or more, of the size and price of the book; and, of equal importance, of the time that the research for it would have required. Scenes from deep time have become such a successful and popular way to present and represent the history of our planet that a narrative and analysis covering their use in the past hundred and thirty years would have needed at least that much time and space, if it were to be more than superficial. But it is not only in terms of pages, illustrations, and research-years that an adequate treatment of later scenes would really need another and different book.

The difference between this book and any potential sequel centers on the concept of a "genre." That invaluable, untranslatable, and still not fully anglicized word lies at the heart of the historical interpretation presented throughout this book, as has been explicit on many of its pages. The argument has been that scenes from deep time were a novel *kind* of picture, developed at a particular period of history in highly specific circumstances. Collectively, the scenes that have been analyzed here (and many others not reproduced) constitute a genre, in the same sense that landscapes and novels and operas are genres. Each genre emerged at a specific historical period; each drew on and adapted preexisting cultural resources of other kinds; each was developed by specific practitioners into a definitive or canonical format. Once developed, however, each proved capable of conveying highly diverse and even incompatible contents in equally varied styles. In other words, implicit in the concept of a genre is a distinction between the history of its *origin* and the subsequent history of its *usage*. Opinions may differ on the historical point at which a genre reached maturity—with, say, Poussin or Constable, Austen or Dickens, Monteverdi or Gluck?—just as they may on the point at which it becomes so radically transformed in the course of practice that it should no longer count as the *same* genre. But nonetheless, there may be some consensus about the exemplars that illustrate its mature arrival on the historical stage.

This book, then, has described and analyzed the origin, emergence, and development to maturity of the genre of scenes from deep time. The book began with a survey of one of the preexisting genres that later served as resources for the creation of a new one, namely, the older tradition of biblical illustration. It then switched to another well-established tradition, that of natural history illustration, and traced its adaptation to deal with fossils and the organisms of which they were the fragmentary remains. That latter tradition led directly into the first true scenes from deep time, defined as "true" because they set many reconstructed organisms into a landscape seen from a tacitly human viewpoint. The nascent genre of such scenes then spread rapidly from specialist reader-viewers to a much wider public. The earlier tradition of biblical illustration was fully assimilated somewhat later, when various isolated scenes were consolidated into a single sequence illustrating the whole history of life on earth. The denizens of the deep past were then released from the limitations of paper, when they were first reconstructed at full size and in three dimensions (thereby arguably initiating a separate offshoot genre). Finally—as far as this book goes—all these exemplars were consolidated into a fuller sequence of scenes, which embodies pictorial conventions that continue to be utilized to the present day.

It is immaterial whether, say, De la Beche or Unger or Figuier is regarded as the "founder" of the genre or as marking its maturity. My claim, and the justification for closing this book with Figuier, is simply that by his time, if not earlier, the genre was well established. By the 1860s it could convey, in terms that a vast public could appreciate, the scientists' vision of the immeasurable expanses of the history of the earth, all the way from the origins of life to the origins of humanity. Conversely, it could also be employed on the popular level for social and even political purposes, with or without the connivance of the scientists. Since that time, it has continued to convey the scientists' vision, vastly enlarged and enriched as it has been by later research, and also to serve wider social purposes. Either way, it has been used to present and represent the world of deep time to an ever-growing public: the public—and not least the *young* public—that now buys popular books and watches television programs on dinosaurs, and throngs our museums to see them in the full glory of their illusory resurrection.

Notes

Introduction

1. The felicitous phrase "deep time" is borrowed from the fine evocation of the world of the modern geologist in John McPhee's *Basin and Range* (1981). By analogy with the "deep space" of the astronomer, it expresses the unimaginable magnitudes of the prehuman or prehistoric time scale.

2. Rudwick, "Visual Language" (1976).

3. For example, Latour and Noblet, *Les "vues" de l'esprit* (1983); Lynch, "Discipline and the Form of Images" (1985); Latour, "Visualisation and Cognition" (1986); Lynch and Woolgar, *Representation in Scientific Practice* (1988); and Fyfe and Law, *Picturing Power* (1988).

4. Hacking, *Representing and Intervening* (1983), p. 138.

5. Rudwick, "Encounters with Adam" (1989).

Chapter One

1. The term "virtual witnesses" is borrowed from Shapin and Schaffer, *Leviathan and the Air Pump* (1985). There it is used to denote the "literary technology" by which an early modern natural philosopher such as Robert Boyle sought to convince his readers of the reality of the phenomena he described, as if they too had been present when the experiment was performed. I extend the term here to denote the similar "making real" of a scene that even the scientist could not really witness. What the two contexts have in common is that success in either case recruits the virtual witnesses as the author's allies in building up the authority of the claims embodied in the textual or visual representation.

2. For classical history there was of course some contemporary pictorial evidence, such as sculptural friezes and painted vases. For biblical history, the iconophobic character of Judaism and, therefore, of early Christianity restricted the evidence to the strictly textual.

3. On the science of chronology, see Wilcox, *Measure of Times Past* (1987), and Grafton, *Defenders of the Text* (1991), chap. 5.

4. See for example Prest, *Garden of Eden* (1981). Allen, *Legend of Noah* (1949), though somewhat comparable in topic, gives little attention to pictorial materials. One kind of illustration that will not be considered here, because it has only a tenuous connection with scenes from deep time, is that in which the whole earth is viewed as if from outer space, rather than from a human, earth-bound viewpoint. Perhaps the best example is the frontispiece of Burnet's *Sacred Theory of the Earth* (1680–89), with its seven globes representing seven successive phases (past, present, and future) of the earth as seen by the cherubim that frame the design; this has been reproduced in, for example, Rudwick, *Meaning of Fossils* (1972, p. 79, fig. 2.7) and Gould, *Time's Arrow* (1988, p. 20, fig. 2.1).

5. The classic treatment of this theme is Ivins, *Prints and Visual Communication* (1953). Copper engravings required more time and skill and were therefore more expensive, but the plate lasted longer and therefore permitted the printing

of a larger number of copies, and the process allowed much more subtle pictorial effects.

6. Scheuchzer, *Herbarium Diluvianum* (1709). See Fischer, *Johann Jakob Scheuchzer* (1973).

7. Scheuchzer, *Home Diluvii testis* (1726). The reinterpretation of the fossil as an amphibian was by Cuvier, in the second edition of his *Recherches sur les ossemens fossiles* (1821–24): see Jahn, "Notes on Dr. Scheuchzer" (1969).

8. Scheuchzer, *Physica sacra* (1731–33); *Kupfer-Bibel* (1731–35); *Physique sacrée* (1732–33). There was also a Dutch edition for the Low Countries, both Protestant and Catholic: *Geestelÿke natuurkunde* (1735–39).

9. See for example Prest, *Garden of Eden* (1981).

10. An important and well-known precedent for Scheuchzer's illustrations of the Flood was Athanasius Kircher's *Arca Noë* (1675), with its detailed analysis of the construction and contents of the Ark, and its double-page engravings of the Flood in progress (e.g., pp. 126–27, 154–55). A selection of Kircher's illustrations is well reproduced in Godwin, *Athanasius Kircher* (1979); more generally, see the classic work by Allen, *Legend of Noah* (1949), particularly the appendix on *Arca Noë*. The frontispiece of Burnet's *Sacred Theory of the Earth* (1680–89) includes a cherubim-eye-view of the whole earth at the time of the Flood, which is clearly related to this pictorial tradition, although its perspective from outer space separates it significantly from the plausibly *human* viewpoint of both Kircher's and Scheuchzer's designs.

11. Buffon, *Époques de la nature* (1778). The standard modern edition is by Roger, "Buffon" (1962), and the work is evaluated in a biographical context in Roger, *Buffon* (1989), chap. 23.

12. See Thackray, "Parkinson's *Organic Remains*" (1976). Morris, *James Parkinson* (1989), gives useful biographical information and reprints the essay on "Parkinson's disease" (pp. 151–75).

13. Rudwick, "Geological Society" (1963); Laudan, "Ideas and Organizations" (1977).

14. Inkster, "London Science" (1979); Miller, "Micropolitics of Science" (1986); Morris, *James Parkinson* (1989), chap. 3.

15. Martin's *oeuvre* as a whole is illustrated in Johnstone, *John Martin* (1974), and analyzed in Feaver, *Art of John Martin* (1975).

16. My interpretation here differs from that offered by Rupke (*Great Chain of History* [1983], pp. 77–78), who sees in the picture the impact of Cuvier's catastrophist geology. It

may well be true that Cuvier saw the painting in Martin's studio during a visit to London, and even that he told Martin he agreed with the astronomical cause of the Deluge alluded to. But I see nothing in Martin's work to suggest a specifically Cuvierian input, or indeed an input from any new research of the period. The animals in Martin's scene, for example, are unmistakably modern lions, elephants, giraffes, etc. (see fig. 11), *not* the extinct mammal species that Cuvier had reconstructed (see chap. 2).

17. Martin, *Illustrations of the Bible* (1838). The *Pictorial Bible* (1836–38) published by Charles Knight, and the Catholic *Bilder-Bibel* (1836), are popular examples of the genre in this period; but neither contains a sequence of scenes of the Creation or the Deluge.

18. Doré, *Sainte Bible* (1866) and *Holy Bible* (1866).

19. See Rappaport, "Borrowed Words" (1982), and Rossi, *Dark Abyss of Time* (1984).

Chapter Two

1. An excellent modern summary of eighteenth-century geology is in Gohau, *History of Geology* (1991); more detail is in his *Sciences de la terre* (1990).

2. Buffon's *Époques de la nature* (1778), another good example, had no illustrations at all. On the paucity of illustrations in eighteenth-century "geological" publications generally, see Rudwick, "Visual Language" (1976).

3. This common supposition may also help explain the rarity of eighteenth-century attempts to reconstruct even individual fossil organisms. One example, notable mainly for its almost unique status in this respect, is the trilobite, reconstructed with imagined appendages, in Schroeter, *Beyträge zur Naturgeschichte* (1774–76), Bd. 1, Tab. 1, Abb. 7. This engraving, the only one of its kind in Schroeter's volumes, is reproduced in Langer, *Paläontologische Buchillustration* (1976), p. 386, *sub* no. 6678.

4. Cuvier's work with fossils is described and analyzed in Coleman, *Georges Cuvier, Zoologist* (1963), chap. 5; Rudwick, *Meaning of Fossils* (1973), chap. 3; and Outram, *Georges Cuvier* (1984), chap. 7.

5. López Piñero, "Juan Bautista Bru" (1988). The engraving was still unpublished when Cuvier first saw it.

6. Cuvier, "Quadrupède inconnue" (1796).

7. Cuvier published a series of articles on these fossil mammals, in the *Annales du Muséum*, between 1804 and 1810, and then reissued them (with other important mate-

rial) in his four-volume *Recherches sur les ossemens fossiles* (1812).

8. Cuvier, *Recherches sur les ossemens fossiles,* 2d ed. (1821–24). The Montmartre mammals are described in vol. 3 (1822).

9. Outram, *Georges Cuvier* (1984), makes clear the precarious and vulnerable character of his career in Napoleonic and Restoration Paris, not least as a Protestant and by origin no Frenchman.

10. Theunissen, "Cuvier's *lois zoologiques*" (1986), rightly emphasizes the marginal role of reconstructions in Cuvier's research, compared to his central concern to assign the fossil bones to their correct place in his taxonomy.

11. Rupke, *Great Chain of History* (1983), pp. 31–41. Buckland's Oxford position was formally that of "Reader" in geology and mineralogy, but he was commonly referred to as "Professor."

12. The *Outlines* (1822) was a major revision by Conybeare of an earlier book by his nominal coauthor (and publisher) William Phillips.

13. There are copies among the papers of several of the leading British geologists; Cuvier's copy is in MS 634(1), Bibliothèque Centrale, Muséum Nationale d'Histoire Naturelle, Paris.

14. The anonymous artist was probably not Conybeare himself: the text too is lithographed, which would have involved the specialized craft skill of writing elegantly in mirror image. Lithography was still a relatively new and untried medium for scientific illustration, although it had been used widely, especially for sheet music, since its invention at the turn of the century. It involves taking paper impressions from a prepared and inked surface of a fine-grained "lithographic stone" (a sheet of zinc was later substituted for the heavy and breakable slabs of stone). It is capable of translating an artist's tonal effects onto multiple paper copies, without the interposition of the engraver's techniques of conventional hatched shading. It was so much better than engraving as a medium for depicting objects such as fossils—and cheaper too—that it was adopted by the Geological Society for the plates in its *Transactions,* not long after Conybeare's cartoon was produced. See Ivins, *Prints and visual Communication* (1953); Twyman, *Lithography* (1970); and, for geological illustration, Rudwick, "Visual Language" (1976).

15. Conybeare and De la Beche, "New Fossil Animal" (1821). De la Beche (see below) provided the stratigraphical background.

16. Conybeare, "Skeleton of the Plesiosaurus" (1824); Cuvier, *Recherches sur les ossemens fossiles,* 2d ed. (1821–24); vol. 5 (1824), chap. 5 (Conybeare's reconstructed skeletons are redrawn as pl. 32).

17. Buckland, "Discovery of Coprolites" (1835) and "New Species of Pterodactyle" (1835), enlarged from a single paper read to the Geological Society in 1829. The contemporary term "pterodactyle" will be used throughout this book, rather than its modern equivalent "pterosaur." For the history of its representations, see Padian, "The Case of the Bat-winged Pterosaur" (1987).

18. This paragraph is based on extensive unpublished research by Hugh Torrens, to whom I am particularly grateful for allowing me to summarize it here. That De la Beche had a local artistic, and perhaps folkloric, tradition to support his design is suggested by the gouache painting "The Lyme Regis Dragon," dated 1829 and attributed to the Reverend G. Howman, now in the Lyme Regis Museum. It shows a winged dragon flying far above a sailing ship in stormy seas off a rocky coast. In pose if not in anatomy, the dragon strongly recalls De la Beche's reptiles. I am much indebted to Jim Secord for giving me a photograph of this painting. For the Oxford connection, see Rupke, *Great Chain of History* (1983), p. 146.

19. The aquarium craze dated from the 1850s, after the seawater aquarium was developed from the similarly self-sustaining "Wardian case" for terrestrial plants; in economic terms it was made possible, at least in Britain, by the repeal in 1845 of the high tax on plate glass. See Allen, *Naturalist in Britain* (1976), pp. 132–40; Rehbock, "Victorian Aquarium" (1980); and Barber, *Heyday of Natural History* (1980), pp. 115–24. A good example of a half-subaqueous view of a diver on a wreck, dated 1832, is reproduced in McKee, *History under the Sea* (1968), p. 9.

20. Rudwick, "Caricature" (1975), reproduces the sequence of De la Beche's sketches that preceded his adoption of a design derived from *Duria antiquior.*

21. Rudwick, "Caricature" (1975); Gould, *Time's Arrow* (1987), pp. 98–104, 137–42. The earlier misidentification of "Professor Ichthyosaurus" as Buckland was due to his son: F. Buckland, *Curiosities* (1857), frontispiece and pp. viii–ix.

22. See Langer, "Georg August Goldfuss" (1971).

23. Goldfuss, "Reptilien der Vorzeit" (1831). It is only a conjecture that Goldfuss designed, and Hohe drew, a scene of pterodactyles *before* they saw De la Beche's *Duria antiquior;* but it is plausible to suppose that if the chronology had been

reversed, their modest reconstruction would have shown some features, or creatures, borrowed from the English scene.

24. Goldfuss, *Petrefacta Germaniae* (1826–44). The illustrations were from drawings by an unnamed artist in the lithographic firm of Arnz in Düsseldorf (preface to first volume).

25. That it was based on De la Beche's design will be more apparent if one or the other is viewed in a mirror, since the lithographic process involves just such a reversal. The print is anonymous, but is likely to have been by Hohe, since he designed both its predecessor (fig. 21) and its successor (fig. 40).

26. An errata slip mentions the late addition of the "übersichtliche Darstellung der Jura-formation" as an extra plate. A copy was received at the Geological Society in London on 30 September 1831.

27. Buckland's "Megalosaurus" (1824) was read to the Geological Society at the same meeting as Conybeare's paper restoring the skeletons of the ichthyosaur and plesiosaur.

28. A similar rendering of these mammals, borrowed either from De la Beche's scene or directly from Cuvier, is the only illustration of its kind in Thomas Brown's edition of Goldsmith's *History of the Earth* (1832), vol. 1, pl. 3.

29. McCartney, *De la Beche* (1977).

Chapter Three

1. Wood engravings are printed from blocks of hard boxwood engraved on the end grain. In the nineteenth century, they were often referred to as "woodcuts," but the technique is quite distinct from that used for the woodcuts that decorated early printed books. Specifically, they were capable of much finer detail and much longer print runs. Although not as subtle a medium as either copper engraving or lithography, wood engravings had the great advantage that they could be printed on the same paper as text, which not only made them more economical but also meant that illustrations could be placed on the same pages as the text to which they referred. See Ivins, *Prints and Visual Communication* (1953).

2. Guérin, *Dictionnaire pittoresque* (1834–39).

3. Guérin, *Dictionnaire pittoresque* (1834), vol. 1, pl. 22 (opp. p. 176), depicts the birds *Crotophaga* ("Ani") and *Plotus* ("Anhinga"), and the snake *Anguis* ("Anguis").

4. Boblaye, "Animaux fossiles" and "Animaux perdus" (1834); Corsi, "French Transformist Ideas" (1978), and *The Age of Lamarck* (1988).

5. Conybeare to Buckland, 4 July 1834, quoted in Howe, Sharpe, and Torrens, *Ichthyosaurs* (1981), p. 22.

6. Buckland, *Geology and Mineralogy* (1836), vol. 2, pl. 1. Some of the fossils are shown as reconstructed individual bodies, for example, minute copies of Cuvier's Tertiary mammals (fig. 16); but there is no attempt to set them in any landscape scene.

7. Kaup, *Mammifères inconnus* (1832–39).

8. Klipstein and Kaup, *Schädel des Dinotherii* (1836).

9. This nice detail, which I had missed, was kindly pointed out to me by Wolfhart Langer: it was first noted in Koenigswald, "Das Dinotherium von Eppelsheim" (1982), a reference for which I am also indebted to Professor Langer. A similar playful whimsy appears on the front of the monograph, above the vignette, where the letters composing the word "Atlas" are formed from fossil bones and the implements of excavation.

10. On the significance of vignettes, see Rosen and Zerner, *Romanticism and Realism* (1984), chap. 3.

11. Buckland, *Geology and Mineralogy*, 2d ed. (1837), pl. 2′, p. 603.

12. According to Roselle, *Samuel Griswold Goodrich* (1968), Goodrich reckoned later in life that he was the author of about 170 books, of which 116 had appeared under the name of Peter Parley, and that their combined sales had been about seven million copies. The British Library catalog gives the author of *Wonders of Earth Sea and Sky* as Samuel Clark, not Samuel Goodrich; the true identity behind the pseudonym in this particular case—it was used by others besides Goodrich—is not important for the present argument.

13. On the long-established moralizing tradition in children's books on natural history, see Ritvo, "Learning from Animals" (1985).

14. Torrens and Cooper, "George Fleming Richardson" (1986), gives valuable biographical information.

15. Delair and Sarjeant, "Earliest Discoveries of Dinosaurs" (1975); Dean, "Gideon Algernon Mantell" (1990). Mantell's manuscript reconstruction of the iguanodon skeleton—in the same style as Cuvier's palaeotherium (fig. 14) but based on far fewer bones—is reproduced in Williams, "Dinosaurs" (1991), fig. 2.

16. Mantell, "Age of Reptiles" (1831).

17. Hawkins, *Book of the Great Sea-Dragons* (1840). Hawkins continued to commission Martin to illustrate some of his later books, such as his long epic poem—dedicated to Queen Victoria—the *Wars of Jehovah* (1844), which includes a vignette of the Deluge (p. 354).

18. Hartmann, *Schöpfungswunder der Unterwelt* (1841),

vol. 2, figs. 61, 60, and 58, respectively; Langer, "Frühe Bilder aus der Vorzeit" (1990).

19. Pictet, *Traité élémentaire de paléontologie* (1844–46), vol. 4 (1846), pls. 19, 20.

20. Pictet's reduced copy of *Duria antiquior* was probably the source for the version in Johann Georg Heck's massive *Bilder-Atlas* (1849, Abt. 1, pl. 1). This was the only scene from deep time in this major collection of visual images for all the sciences in mid-century.

21. Milner, *Gallery of Nature* (1846), p. 611. Later in the book are small and derivative versions of the commonest scenes in popular books, namely those of "Saurians" and the "Paris basin" (pp. 724, 745).

22. See Altick, *Shows of London* (1978), particularly chap. 10; and Schivelbusch, *Railway Journey* (1989), chap. 4, and particularly the series of scenes from the Paris-Orleans line, published in 1843 (p. 64).

Chapter Four

1. Unger, *Chloris protogaea* (1841–47). Reyer, *Franz Unger* (1871), is the standard source of biographical information.

2. Unger, *Urwelt* (1851). The work bears no date, though the preface is dated 18 June 1847. However, Reyer (*Franz Unger* [1871], p. 47) describes how publication was delayed until 1851, by which time Unger himself had moved from Graz to Vienna. The second edition (Leipzig, 1858), with two extra plates, is described later in this book (chap. 5). All the plates are reproduced here, although for practical reasons the scenes have had to be greatly reduced in size. Since the book is now a rarity in all its editions, including the probably pirated editions in English (see chap. 5), and since it is so important for the theme of this book, Unger's texts are given in full, although they are longer than most of the other texts reproduced here.

3. The *Bilder-Bibel* (1836), unlike Scheuchzer's earlier work, jumps pictorially straight from God's original creative act to the Garden of Eden, omitting any scenes of the successive "days" of Creation. Nonetheless, it did continue the tradition of portraying thereafter a temporal *sequence* of scenes for the rest of biblical history.

4. See Rudwick, "Uniformity and Progression" (1971); Bowler, *Fossils and Progress* (1976). It was this consensus that Lyell's "steady-state" model, expounded in his *Principles of Geology* (1830–33), had signally failed to undermine.

5. Unger equates this, in his French text, with the much older (pre–Coal Measures) *vieux grès rouge,* or "Old Red Sandstone" of Britain and other parts of northern Europe. The later English edition of Unger's work (*Primitive World,* 1855) corrects this by correlating the *Totliegende* with the "Permian" group that Murchison had proposed in 1841, based on formations in the region of Perm, west of the Urals.

6. Owen, "Genus *Labyrinthodon*" (1841); "British Fossil Reptiles" (1842).

7. Curiously, Unger's text (41) alludes also to the distinctly different reptile *Hylaeosaurus,* although in the plate (fig. 51) all three individuals clearly have the characteristic rhinoceroslike horn of the iguanodon. Unger's caution about the form of the iguanodon was vindicated much later in the century, when the discovery of almost complete skeletons revealed that it had been a bipedal animal quite different from Mantell's giant lizard.

8. Agassiz, *Études sur les glaciers* (1840); see Carozzi's introduction to his English edition, Agassiz, *Studies on Glaciers* (1967).

9. See for example Grayson, *Human Antiquity* (1983), chap. 6.

10. The subscribers are listed at the end of the text. According to Kirchheimer, "Einführung der Photographie" (1982), the original price was 16 Thaler.

11. On the traditional "invisibility" of scientific technicians and craftsmen, see Shapin, "The Invisible Technician" (1989).

Chapter Five

1. Zimmermann, *Wunder der Urwelt* (7th ed., 1855), also issued as vol. 3, part 1, of his *Physikalische Geographie* (5th ed., 1855–58). I have not seen the earlier editions, and the frontispiece may have been added in 1855. The work was translated as *Le monde avant la création de l'homme* (1857) and, in greatly abridged form (without the frontispiece), as *Wonders of the Primitive World* (1869). Its revision by the French astronomer and popularizer Camille Flammarion (1885) made it a major vehicle of popular geology in the later part of the century.

2. Brewer, *Theology in Science* (1860), frontispiece, steel engraving by G. Whymper. Among other examples are the small scenes, derived from De la Beche's a quarter-century earlier (figs. 23–25), in the textbook *Geology, Mineralogy and Crystallography* by David Ansted, James Tennant, and Walter

Mitchell (1855); and those in the *Illustrated Natural History* (1859) published in New York by Samuel Goodrich (under his own name rather than as "Peter Parley"), which are clearly derived from Milner's vignette (fig. 41) and from Martin's scenes for Mantell and Thomas Hawkins (figs. 35, 36).

3. Livingstone, "Preadamites" (1986), describes the nineteenth-century movement and its twentieth-century sequel, but does not mention Duncan's work. For the Pre-Adamitism of the seventeenth century, see Popkin, *Isaac La Peyrère* (1987), and Grafton, *Defenders of the Text* (1991), chap. 8.

4. See for example Altick, *Shows of London* (1978), chap. 34.

5. Owen, "British Fossil Reptiles" (1842). The theoretical inferences are in the "Summary," pp. 191–204.

6. Desmond, "Designing the Dinosaur" (1979); and see, more generally, his *Politics of Evolution* (1989).

7. The large-scale map in Owen's guidebook (*Ancient World*, 1854) shows a *second* island (which still exists) with unidentifiable animals on it, in the right position to have been the planned site for Tertiary mammals; but the text mentions only the reptiles on the Secondary island. An undated revised version of Hawkin's sketch (fig. 63) suggests that two Tertiary mammals, an "Irish elk" and a mylodon, were in fact added later. But they were placed not on a separate island but *behind* the Secondary reptiles, thus destroying the chronological sequence of the original display. (A modern photograph of this later version of figure 63 is in the folder of "Waterhouse Hawkins drawings, photographs etc." in the Hawkins papers at the Natural History Museum, London.)

8. A watercolor sketch among the Hawkins papers (see above, note 7) shows the strata as a series of overlapping, gently sloping rock platforms, with each reptile model standing directly on its appropriate formation. This probably represents an earlier design that was discarded; it would have integrated the animals with their respective strata even more effectively.

9. They still stand on the original island, in virtually the same positions as in figure 63. Desmond, "Fragile Dinosaurs" (1974), includes his fine modern photographs of some of them, newly repaired and repainted. Their impact is perhaps even greater than in the 1850s, since the terrestrial forms now lurk in thick vegetation among mature trees. Appropriately, modern visitors are guided to the site by signs reading "Monsters."

10. W. Hawkins, "Visual Education" (1854), p. 444.

11. In one respect the models were more vivid than the living animals: the fossil reptiles were shown in what purported to be a naturalistic setting, rather than being seen behind bars in cages or "houses," as they were at the zoo. See Blunt, *Ark in the Park* (1976); Ritvo, *Animal Estate* (1987), chap. 5.

12. Unger, *Primitive World* (1855). The book is undated, but the British Library's copy was acquired under the Copyright Act on 1 August 1855. It can hardly be earlier than 1854, when Hawkins's exhibit was opened. Highley's "Publisher's Preface" (pp. 1–2) explains that his plates are "albuminized paper positives" from "collodion reductions" of Kuwasseg's originals, (the photographs measure only 166 × 133 mm). The book was intended as the first in the series "Photography in its Application to Palaeontology," which would otherwise be devoted to pictures of fossil specimens. The price was two guineas (in modern notation, £2.10): it was not a cheap book.

13. Highley later published a second edition of his English translation, incorporating photographs by Russell Sedgfield of Unger's two new plates (at the still further reduced size of 105 × 73 mm). Like the first, this edition is undated, but the British Library copy shows the acquisition date as 18 November 1864. It was limited to 250 copies, in order to maintain the quality of the photographs. Highley also added as a frontispiece a photograph of a set of Hawkins's small-scale models, supplied and arranged by Hawkins on an irregular rock plinth in the style of an elaborate Victorian sculpture. These were explicitly a substitute for photographs of the full-sized versions, which the Crystal Palace Company's official photographers declined to supply: here at least the enterprising Highley was foiled by copyright restrictions!

14. Bowler, *Non-Darwinian Revolution* (1988), argues persuasively that Darwin's theory must be understood in the context of a wide range of other evolutionary speculations.

15. Tennant, "Waterhouse Hawkins's Restorations," pamphlet dated July 1860, advertising Hawkins's "Struggles of Life" at the price of 12 shillings. I have not located a copy of the lithograph itself.

16. The pamphlet cited in note 15 also offered sets of models of Hawkins's Crystal Palace reptiles, reduced to the scale of 1:12. They were, predictably, of a pterodactyle, an iguanodon, a megalosaur, two plesiosaurs, an ichthyosaur, and a labyrinthodon, and were said to be modeled "in strict accordance with the criticism and sanction of the highest sci-

entific authorities." There is a photograph of the iguanodon model in Czerkas and Olson, *Dinosaurs Past and Present* (1987), vol. 1, p. xiv.

17. Hawkins, "Extinct Animals," set of six "double-tinted" lithographs, 40 in. by 29 in. Figs. 70–75 are reproduced from the set in the Hawkins papers at the Natural History Museum, London. I am very grateful to Jim Secord for alerting me to their existence there. They are undated, but are advertised (in a way that suggests they were not new) in a later version of Tennant's pamphlet on "Struggles of Life." This too is undated, but it refers to Lyell's *Antiquity of Man* (1863) as his "new work." This suggests that Hawkins's "Extinct Animals" set was published between 1861 and 1863. The charts were priced at 6 shillings each, or £1 10s. for the set, and were said to be "well adapted for the Educational purposes for which they are intended" (a copy of this later Tennant pamphlet is also in the Hawkins papers).

18. An engraving of Hawkins's workshop in New York, c. 1870, is reproduced in Czerkas and Olson, *Dinosaurs Past and Present* (1987), vol. 1, p. xvi, and shows the first American dinosaur. Desmond, "Fragile Dinosaurs" (1974) describes Hawkins's American work and the political machinations that brought his New York project to a halt.

19. Boucher's work is described and evaluated in Grayson, *Human Antiquity* (1983), chaps. 8, 9, and in Cohen and Hublin, *Boucher de Perthes* (1989). See also Laurent, "Origine de l'homme" (1989).

20. Boitard, *Paris avant les hommes* (1861), fig. opp. p. 10: "M. Boitard et le diable boiteux sur un aerolithe."

Chapter Six

1. Figuier, *Histoire des merveilleux* (1860). Cardot's introduction to her edition of Figuier's work on electricity (*Merveilles de l'électricité*, 1985) gives useful biographical background.

2. Figuier, *La terre avant de Déluge* (1863). Although the English editions (see below) translated the title as "The *world* before the Deluge," I render it here by the more accurate term "earth": in both languages the difference is significant.

3. d'Orbigny, *Cours élémentaire* (1849–52).

4. Figuier's book was to be the first in a series entitled "Tableau de la nature: Ouvrage illustré à l'usage de la jeunesse." See also text 86 below.

5. See Verne, *Cinq semaines en ballon* (1865): for example,

Riou's drawing of the African elephant snared by the adventurers in their balloon ("L'animal essayait vainement de se débarrasser," opp. p. 136) is unmistakably similar in style to his pictures of extinct animals in Figuier's book.

6. See the dramatic engraving of its original discovery at Maastricht, published in Faujas de Saint-Fond, *Montagne de Saint-Pierre* (1799); this is reproduced in Rudwick, *Meaning of Fossils* (1972), fig. 3.7, p. 128, and in Laurent, *Paléontologie et évolution* (1987), fig. 5, p. 155.

7. Grayson, *Human Antiquity* (1983), chap. 9. The significance of the "Moulin-Quignon jaw" was in fact controversial from the first, but its discovery did mark a decisive swing in expert opinion about the larger issue.

8. The first English edition was translated by W. S. Ormerod and dedicated to Murchison; the sales figure is given in his "Envoi." The revised translation was by Henry William Bristow (1817–89).

9. The agreement over Riou's engravings must have been made some time before Fraas's book appeared, because it reproduces the earlier version of some of the pairs that have been noted above in the history of Figuier's book. Riou's illustrations are slightly superior in quality in Fraas's book, because they were printed on better paper.

10. At least this was the intention indicated in the "Benachrightigung für den Buchbinder," which gave the positions in the text at which the plates were to be inserted when the paperback installments were assembled. Of course this involved some confusion between the two deluges that Figuier had distinguished, since the one depicted in Fraas's frontispiece was clearly *not* the one that separated the human world from the prehuman. Perhaps for this reason, Fraas retitled Riou's scene of the prehuman "Déluge du nord de l'Europe" (fig. 100) as the "Europaeische Eiszeit," while the preceding scene of a periglacial Quaternary landscape (fig. 99) became a "Landschaft der Mammuth Zeit."

Chapter Seven

1. For later nineteenth century "scenes", see for example the immensely popular Flammarion, *La monde avant la création de l'homme* (1886); for the early twentieth, Abel, *Tierwelt der Vorzeit* (1922), and *Rekonstruktion vorzeitlicher Wirbeltiere* (1925); and for more recent decades, Czerkas and Olson, *Dinosaurs Past and Present* (1987). It is no accident that scenes of *dinosaurs* are those that best exemplify the modern genre.

2. For the history of ichthyosaur reconstructions, including the twentieth century, see Howe, Sharpe, and Torrens, *Ichthyosaurs* (1981).

3. The notion of a "cascade" is introduced by Latour, "Visualisation and Cognition" (1986), though not in the context of a rational reconstruction.

4. On the importance of replacing the language of "influence" in the history of art with terms that make clear the actors' *choice* of particular earlier models as their *resources*, see, for example, Baxandall, *Patterns of Intention* (1985), pp. 58–62. The same argument applies equally to the history of science, and to the specific case of scenes from deep time, which lies on the border between art and science.

5. See, respectively, Secord, *Victorian Geology* (1986), and Rudwick, *Devonian Controversy* (1985).

6. The tradition of such illustrated travel accounts is described and analyzed in Stafford, *Voyage into Substance* (1984).

7. The frontispiece of Burnet's *Sacred Theory of the Earth* (1680–89) is reproduced in, for example, Rudwick, *Meaning of Fossils* (1972, p. 79), and Gould, *Time's Arrow* (1988, p. 20).

8. For example, Gainsborough haughtily declined to paint "*real Views* from Nature in this Country [i.e., England," and Fuseli referred to topography as "the last branch of uninteresting subjects" for the artist: quoted respectively in Herrmann, *British Landscape Painting* (1973, pp. 39–40, and Alfrey, "Ordnance Survey" (1990), p. 23. See also Twyman, *Lithography* (1970), p. 12, on the conflict between the aims of topography and the taste for the "picturesque." The political interpretation of landscape art in this period, developed by Bermingham, *Landscape and Ideology* (1986), is hardly applicable to scenes from deep time. The distinction drawn by Alpers, *Art of Describing* (1983), between "descriptive" and "Albertian," or narrative, modes in the history of art, although illuminating for the comparison between Dutch and Italian painting in the early modern period, is likewise difficult to apply to scenes from deep time. The objectives of these scenes were clearly "descriptive" in the sense of claiming to represent, with all possible accuracy and detail, what the time-traveling human eye would have seen—"ideally"—at some point in the deep past. But they also had "narrative" objectives, in the sense of claiming to represent episodes in a long history of life on earth, each of which is to be "read" as part of a story: hence the windowlike frames and keyed captions, which invited responses such as "there's an ichthyosaur eating a plesiosaur!" in parallel to "there's Methuselah awaiting his fate in the Deluge!"

9. Bryson, *Word and Image* (1981).

10. Shapin, "The Invisible Technician" (1989).

11. The illustrations in Gould's *Wonderful Life* (1989) make this point well. The superb pictorial reconstructions of the bodies of the bizarre Burgess Shale animals (figs. 3.12, 3.18, 3.21 etc.) were commissioned specially for this deservedly popular book; the reconstructions published in the specialist scientific literature (and also reproduced in Gould's book) are far more schematic. Likewise the "realistic" scenes of the Middle Cambrian period—giving an aquarium–like view—are reproduced from museum displays designed for the general public (figs. 1.1, 1.2); whereas the nearest approach to "scenes" in the literature aimed at scientists are block-diagrams depicting the inferred habitats of the various genera in highly schematic form (figs. 3.62, 3.65).

12. Shapin, "Science and the Public" (1989), gives a useful review of recent historical interpretations of the so-called popularization of science.

13. See for example Michael Baxandall's classic *Painting and Experience* (1972).

14. On the popular taste for natural history in Victorian Britain, see Allen, *Naturalist in Britain* (1976), and, for some of its literature, Merrill, *Victorian Natural History* (1989). The popular symbolic and rhetorical meanings attached to wild and domestic animals are explored (likewise only for Britain) in Ritvo, *Animal Estate* (1987).

15. See for example Bann, *Clothing of Clio* (1984), chap. 3, on the early nineteenth-century Romantic taste for historical pictures and museums.

16. Ritvo, *Animal Estate* (1987), makes this point for *living* animals.

17. See Jordanova, "Objects of Knowledge" (1989), and Sorensen, "Theme Parks and Time Machines" (1989).

18. See, for example, the analysis of the early twentieth-century dioramas of African wildlife at the American Museum of Natural History in New York, in Haraway, *Primate Visions* (1989), chap. 3. Characteristically, however, this insightful verbal analysis contains only one photographic reproduction of the visual displays that are its ostensible subject matter—and even that is a close-up of a particular animal (fig. 3.1, p. 32), rather than a complete diorama. Such is the power of the nonvisual (or even anti-visual?) tradition that dominates social and historical studies of science, as Fyfe and Law rightly comment in their editorial introduction to *Picturing Power* (1988).

Sources for Figures and Texts

Figures

1. Scheuchzer, *Physica sacra* (1731), Tab. 6; Genesis 1:9–10.

2. Scheuchzer, *Physica sacra* (1731), Tab. 8; Genesis 1:11–13.

3. Scheuchzer, *Physica sacra* (1731), Tab. 15; Genesis 1:21.

4. Scheuchzer, *Physica sacra* (1731), Tab. 20.

5. Scheuchzer, *Physica sacra* (1731), Tab. 22; Genesis 1:24–25.

6. Scheuchzer, *Physica sacra* (1731), Tab. 23; Genesis 1:26–27.

7. Scheuchzer, *Physica sacra* (1731), Tab. 43; Genesis 7:11.

8. Scheuchzer, *Herbarium Diluvianum* (1709), vignette on title page.

9. Parkinson, *Organic Remains* (1804–11), vol. 1 (1804), frontispiece.

10, 11. Martin, *The Deluge* (1828): the mezzotint after his original painting (now in the Tate Gallery, London).

12. Georges Cuvier's copy of Bru's original print, MS 634 (2), Biblothèque Centrale, Muséum Nationale d'Histoire Naturelle, Paris. Published later in Bru, "Descripcion del esqueleto" (1796), pl. 1. Crude reproduced copies are in Cuvier, "Quadrupède trouvé au Paraguay" (1796), and *Recherches sur les ossemens fossiles* (1812), vol. 4, "Megatherium," pl. 1, fig. 1. On the history of Bru's work, see Lopez Piñero, "Juan Bautista Bru" (1988).

13. Cuvier, "Sur le grande Mastodonte" (1806); reissued in *Recherches sur les ossemens fossiles* (1812), vol. 2, art. 10, pl. 5.

14. Cuvier, "Pierre à plâtre" (1804–8); reissued in *Recherches sur les ossemens fossiles* (1812), vol. 3, mem. 7, unnumbered plate.

15. MS drawing, undated, in Cuvier's hand (MS 635, Bibliothèque Centrale, Muséum National d'Histoire Naturelle, Paris). This is one of a set of three drawings in the same style (that of *Palaeotherium minus* is reproduced on a small scale in Coleman, *Georges Cuvier* [1963], p. 122; that of *P. magnum* is missing). The corresponding reconstructions of the skeletons were published in *Recherches sur les ossemens fossiles* (1812), vol. 3, mem. 7, 3 unnumbered plates.

16. Cuvier, *Recherches sur les ossemens fossiles,* 2d ed. (1821–24), vol. 3 (1822), pl. 66.

17. "The Hyaenas' Den at Kirkdale," anonymous and undated lithographed broadsheet attributed to William Conybeare (c. 1822); see note on text 10.

18. Conybeare, "Skeleton of the Plesiosaurus" (1824), pl. 49.

19. *Duria antiquior:* original lithograph by George Scharf, printed by Hullmandel, based on De la Beche's hand-drawn, water-colored sketch in the De la Beche MSS, National Museum of Wales, Cardiff. The museum has published a full-sized reproduction of the sketch, and it also appears on the cover of Howe, Sharpe, and Torrens *Ichthyosaurs* (1981). What is probably a second version of Scharf's lithograph, with six of the animals numbered and identified in the caption, has been reproduced in McCartney, *Henry De la Beche* (1977),

p. 45; Secord, "Geological Survey" (1986), p. 242; and Rudwick, "Encounters with Adam" (1989), p. 242. Thirty years later, a crudely redrawn engraving (with ten numbers) formed the frontispiece of Francis Buckland's *Curiosities of Natural History* (1860), and has been reproduced in Browne, *Secular Ark* (1983), p. 100. I am indebted for this summary to unpublished research by Hugh Torrens.

20. De la Beche, "Awful Changes" (1830): lithographed broadsheet. This exists in two versions, differing slightly in design and caption. The undated version is almost certainly the original; it is cruder in many details than the version reproduced here, in which the drawing has been improved, and the date "1830" added to De la Beche's signature in the bottom right corner. I am grateful to Doug Bassett and Mike Bassett for clarifying this important point. The varied series of preliminary sketches for this cartoon is analyzed in Rudwick, "Caricature" (1975). I am now, however, convinced that it should be dated 1830, not 1831; the difference is significant in relation to both Lyell's and De la Beche's work around this time.

21. Goldfuss, "Reptilien der Vorzeit" (1831), Taf. 9; also reproduced in Langer, "Früher Bilder aus der Vorzeit" (1990).

22. Goldfuss, *Petrifacta Germaniae* (1826–44), Theil 1, Lieferung 3 (1831), unnumbered plate.

23. De la Beche, *Geological Manual*, 2d ed. (1832), fig. 37, p. 231; wood engraving, from a drawing almost certainly by De la Beche himself. This and the following illustrations do not appear in the first edition (1831), but were reproduced in the third (1833). They were also reproduced in the French translation (1833), but not in the German (1832), although both were based on the second edition.

24. De la Beche, *Geological Manual*, 2d ed. (1832), fig. 79. The quotation is from a footnote on p. 383.

25. De la Beche, *Geological Manual*, 2d ed. (1832), fig. 80. The quotation is from a footnote on p. 385.

26. Phillips, "Organic Remains Restored" (1833).

27. Guérin, *Dictionnaire pittoresque* (1834–39), vol. 1 (1834), pl. 24.

28. T. Hawkins, *Memoirs of Ichthyosauri and Plesiosauri* (1834), frontispiece: lithograph (390 × 265 mm).

29. Buckland, *Geology and Mineralogy* (1836), vol. 2, pl. 22, fig. P, p. 34.

30. Klipstein and Kaup, *Schädel des Dinotherii* (1836): vignette on front cover of folio "Atlas." Langer, "Frühe Bilder aus der Vorzeit" (1990), identifies the artists as Rudolf Hof-

mann (1820–82) of Darmstadt, later a painter and museum director (although at this time still a teenager); and Ludwig Becker (1808–61), who had been working as an artist for Kaup since 1826, and who later emigrated to Australia and perished on the ill-fated Burke and Willis expedition.

31. Klipstein and Kaup, *Schädel des Dinotherii* (1836): vignette on back cover of folio "Atlas," lithograph by "H. & B.," i.e., Hofmann and Becker.

32. Parley, *Wonders of Earth Sea and Sky* [1837], anonymous lithograph opp. p. 5.

33. Parley, *Wonders of Earth Sea and Sky* [1837], anonymous lithograph opp. p. 21.

34. Richardson, *Sketches in Prose and Verse* (1838), frontispiece. The artist, George Nibbs, was probably the father of the better known engraver Richard Henry Nibbs (1816–93), who was active in Brighton later in the century.

35. Mantell, *Wonders of Geology* (1838), frontispiece. The only other reconstruction in the whole book is a small and stylized wood engraving, "The Flora of the Carboniferous Epoch" (p. 581), by Mantell's daughter Ellen Maria, showing one plant of each of nine species, on a low shore. Martin's frontispiece was used in the German edition, *Phänomene der Geologie* (1839), which must have made it widely known in central Europe; it also continued to be used in English editions, even in the posthumous eighth edition (1864).

36. T. Hawkins, *Book of the Great Sea-Dragons* (1840), frontispiece, mezzotint (293 × 198 mm). The caption is taken from the List of Plates.

37. Richardson, *Geology for Beginners* (1842), frontispiece.

38. Richardson, *Geology for Beginners*, 2d ed. (1843), frontispiece.

39. Trimmer, *Practical Geology* (1841), frontispiece.

40. Goldfuss, *Petrefacta Germaniae* (1831–44), Theil 3, Lieferung 8 (1844), pl. 200, bound as frontispiece to the third volume.

41. Milner, *Gallery of Nature* (1846), p. 611.

42. Steel engraving by John Emslie. This rare item is reproduced from a copy in the possession of Mrs. R. A. Gordon, formerly on loan to the Devon County Record Office. I am greatly indebted to Hugh Torrens for telling me of its existence there.

43. Unger, *Die Urwelt in ihren verschiedenen Bildungsperioden* (1851), Taf. 1: lithograph (540 × 310 mm).

44. Unger, *Urwelt* (1851), Taf. 2.

45. Unger, *Urwelt* (1851), Taf. 3.

46. Unger, *Urwelt* (1851), Taf. 4.
47. Unger, *Urwelt* (1851), Taf. 5.
48. Unger, *Urwelt* (1851), Taf. 6.
49. Unger, *Urwelt* (1851), Taf. 7.
50. Unger, *Urwelt* (1851), Taf. 8.
51. Unger, *Urwelt* (1851), Taf. 9.
52. Unger, *Urwelt* (1851), Taf. 10.
53. Unger, *Urwelt* (1851), Taf. 11.
54. Unger, *Urwelt* (1851), Taf. 12.
55. Unger, *Urwelt* (1851), Taf. 13.
56. Unger, *Urwelt* (1851), Taf. 14.
57. This is reproduced from the French edition: Zimmermann, *La monde avant la création de l'homme* (1857), frontispiece.
58. Brewer, *Theology in Science* (1860), frontispiece.
59. Duncan, *Pre-Adamite Man* (1860), folding plate opp. p. 220, lithograph (344 × 190 mm) dated 1859.
60. *Illustrated London News*, vol. 23, wood engraving on p. 600, issue of 31 December 1853.
61. *Illustrated London News*, vol. 24, wood engraving on p. 22, issue of 7 January 1854.
62. Owen, *Ancient World* (1854), vignette on p. 5.
63. W. Hawkins, "Visual Education" (1854), p. 446.
64. "Punch's Almanack for 1855," *Punch* (1855), vol. 28, p. [8].
65. *Punch*, vol. 28, wood engraving on p. 50, issue of 3 February 1855. The ghostly band accompanying the nightmare may allude to another of the entertainments on offer at the Crystal Palace.
66. Buckland, *Geology and Mineraology* (1858), lithograph, pl. 23.
67. Unger, *Urwelt*, 2d ed. (1858), Taf. A.
68. Unger, *Urwelt*, 2d ed. (1858), Taf. B.
69. Tennant, "Waterhouse Hawkins's Restorations," dated July 1860. I have not been able to locate a copy of the lithograph itself; it measured 34 inches by 28 inches, and was priced at 12 shillings.
70. Hawkins, "Extinct Animals," sheet 1. The full caption reads: "Class Reptilia—Enaliosauria, or Marine Lizards [etc.]."
71. Hawkins, "Extinct Animals," sheet 2.
72. Hawkins, "Extinct Animals," sheet 3.
73. Hawkins, "Extinct Animals," sheet 4.
74. Hawkins, "Extinct Animals," sheet 5.
75. Hawkins, "Extinct Animals," sheet 6.
76. Boitard, *Paris avant les hommes* (1861), fig. opp. p. 65.

77. Boitard, *Paris avant les hommes* (1861), frontispiece.
78. Boitard, *Paris avant les hommes* (1861), fig. opp. p. 239.
79. Figuier, *La terre avant le Déluge* (1863), fig. 26.
80. Figuier, *La terre avant le Déluge* (1863), fig. 27.
81. Figuier, *La terre avant le Déluge* (1863), fig. 38.
82. Figuier, *La terre avant le Déluge* (1863), fig. 62.
83. Figuier, *La terre avant le Déluge*, 4th ed. (1865), fig. 69.
84. Figuier, *La terre avant le Déluge*, 4th ed. (1865), fig. 84. This replaced figure 79 of the first edition (1863).
85. Figuier, *La terre avant le Déluge* (1863), fig. 83.
86. Figuier, *La terre avant le Déluge* (1863), fig. 104. The caption is translated from the second edition; in the first it was misprinted.
87. Figuier, *La terre avant le Déluge* (1863), fig. 105. The caption is translated from the second edition; in the first it was misprinted.
88. Figuier, *La terre avant le Déluge* (1863), fig. 131.
89. Figuier, *La terre avant le Déluge* (1863), fig. 132.
90. Figuier, *La terre avant le Déluge* (1863), fig. 182. The caption is translated from the fourth edition (1865, fig. 157), after the following two scenes had been added.
91. Figuier, *La terre avant le Déluge*, 4th ed. (1865), fig. 160.
92. Figuier, *La terre avant le Déluge*, 4th ed. (1865), fig. 186.
93. Figuier, *La terre avant le Déluge* (1863), fig. 189.
94. Figuier, *La terre avant le Déluge* (1863), fig. 240.
95. Figuier, *La terre avant le Déluge* (1863), fig. 263.
96. Figuier, *La terre avant le Déluge* (1863), fig. 280.
97. Figuier, *La terre avant le Déluge* (1863), fig. 294. In later editions, after the corresponding scene of South American life in the Pliocene (not reproduced here) had been redated as Quaternary, "a European landscape" was changed to "the Earth."
98. Figuier, *La terre avant le Déluge*, 6th ed. (1867), fig. 314. In the fourth edition (1865, fig. 296), a similar but inferior engraving, not by Riou, showed the same animals in a different composition.
99. Figuier, *La terre avant le Déluge* (1863), fig. 303. The parenthetical "(Europe)" was in fact only added when, in later editions, the scene of South American life was moved from Pliocene to Quaternary (fig. 98).
100. Figuier, *La terre avant le Déluge* (1863), fig. 304.
101. Figuier, *La terre avant le Déluge* (1863), fig. 310. This was retained in the fourth edition (1865, fig. 301), but by the

sixth edition (1867) had been replaced (see fig. 102 here).

102. Figuier, *La terre avant le Déluge*, 6th ed. (1867), fig. 322.

103. Figuier, *La terre avant le Déluge* (1863), fig. 312.

104. Fraas, *Sündfluth!* (1866). The book was issued in eleven installments, all with the same paper cover.

105. *Punch*, vol. 55, p. 272, issue of 26 December 1868. The distinctive "DM" was used by George du Maurier: see de Maré, *Victorian Illustrators* (1980), e.g., p. 144.

Texts

1. Scheuchzer, *Physique sacrée* (1732), p. 16.
2. Scheuchzer, *Physique sacrée* (1732), p. 25.
3. Scheuchzer, *Physique sacrée* (1732), p. 28.
4. Scheuchzer, *Physique sacrée* (1732), p. 29.
5. Scheuchzer, *Physique sacrée* (1732), pp. 58–59. The quotation is from *Histoire de l'Académie des Sciences* (1710), p. 22 (*medailles* is rendered here as "coins").
6. Parkinson, *Organic Remains* (1804–11), vol. 1 (1804), pp. 13–14.
7. Martin, *A Descriptive Catalogue of the Engraving of the Deluge* (1828), pp. 3, 8.
8. Cuvier, *Recherches sur les ossemens fossiles*, 2d ed. (1821–24), vol. 3 (1822), pp. 244–51.
9. Buckland, "An Assemblage of Fossil Teeth and Bones" (1822), pp. 186–190, 192–93, 195–98, 202, 208; reprinted with minor alterations in Buckland, *Reliquiae Diluvianae* (1823), pp. 19–24, 27–28, 30–37, 42–44, 51.
10. [William Conybeare,] "The Hyaena's Den at Kirkdale" [1822]. The poem is reprinted (with minor amendments or errors) in Daubeny, *Fugitive Poems* (1869), pp. 92–94, where it is attributed to Conybeare and dated 1822. Daubeny was Buckland's colleague at Oxford and was unlikely to have been mistaken in this attribution. The elegant lettering in which this poem is lithographed is certainly the work of a professional draftsman; this makes it likely that the accompanying lithographed cartoon (fig. 17) was also drawn by a professional, though doubtless to Conybeare's instructions.
11. Conybeare to De la Beche, 4 March 1824 (De la Beche MSS, National Museum of Wales, Cardiff); quoted in McCartney, *De la Beche* (1977), p. 44.
12. Conybeare, "Skeleton of the Plesiosaurus" (1824), pp. 388–89.
13. Lyell, *Principles of Geology* (1830–33), vol. 1 (1830), p. 123.

14. Goldfuss, "Reptilien der Vorzeit" (1831), pp. 63–64, 105.
15. Buckland to De la Beche, 14 October 1831 (De la Beche MSS, National Museum of Wales, Cardiff); quoted in Rudwick, "Encounters with Adam" (1989), pp. 241–43.
16. De la Beche, *Geological Manual*, 2d ed. (1832), pp. 383–85. Neither this passage, nor the illustrations, appear in the first edition (1831); but they are reprinted in the third (1833, pp. 343–44) and in the French edition (*Manuel géologique* [1833], pp. 462–64); the latter would have made them widely known on the Continent.
17. Guérin, *Dictionnaire pittoresque* (1834–39), vol. 1 (1834), pp. 193–94.
18. T. Hawkins, *Memoirs of Ichthyosauri and Plesiosauri* (1834), title page and pp. 5, 51.
19. Buckland, *Geology and Mineralogy* (1836), vol. 1, pp. 223–25. quoting in part from "New Species of Pterodactyle" (1829), pp. 217–19. The quotation is from *Paradise lost,* book 2, lines 947–50.
20. Buckland, *Geology and Mineralogy* (1836), vol. 1, pp. 137–38.
21. Parley, *Wonders of Earth Sea and Sky* [1837], pp. 5, 14–20.
22. Parley, *Wonders* [1837], pp. 21–26.
23. Richardson, *Sketches in Prose and Verse* (1838), pp. 6–7, 11.
24. Mantell, journal entry for 27 September 1834, printed in Curwen, *Journal of Gideon Mantell* (1940), p. 125.
25. Mantell, *Wonders of Geology* (1838), vol. 1, pp. 368–69.
26. T. Hawkins, *Book of the Great Sea-Dragons* (1840), pp. 27, 18.
27. Richardson, *Geology for Beginners*, 2d ed. (1843), p. xiii.
28. Trimmer, *Practical Geology* (1841), pp. xxv, xxvi.
29. Goldfuss, *Petrefacta Germaniae* (1826–44), Theil 3, Lieferung 8 (1844), pp. 123–24. The generic names given in brackets in this translation are listed in the original text as footnotes, with both generic and specific names, and with numbers to key them to the illustration. The numbers are omitted here, because many are obscure even on the original print, and some cannot be located at all.
30. Reynolds, "Popular Geology," broadsheet dated 1 October 1849.
31. Unger, *Ideal Views of the Primitive World* (Samuel Highley's translation of *Urwelt*, 1855), Publisher's Preface, pp. 1–2. All the texts in this chapter are reproduced, with minor corrections, from Highley's English translation, which

is clearly based on Schimper's French translation of Unger's original text. However, this English version does preserve an appropriate period flavor, and its differences from the original are more stylistic than substantive. The names of taxonomic authors, appended to specific names of fossils, have been omitted here.

32. Unger, *Ideal Views* (1855), pp. 5–8.

33. Unger, *Ideal Views* (1855). This and subsequent texts are unpaginated.

34. Unger, *Ideal Views* (1855).

35. Unger, *Ideal Views* (1855).

36. Unger, *Ideal Views* (1855).

37. Unger, *Ideal Views* (1855).

38. Unger, *Ideal Views* (1855).

39. Unger, *Ideal Views* (1855).

40. Unger, *Ideal Views* (1855).

41. Unger, *Ideal Views* (1855).

42. Unger, *Ideal Views* (1855).

43. Unger, *Ideal Views* (1855).

44. Unger, *Ideal Views* (1855).

45. Unger, *Ideal Views* (1855).

46. Unger, *Ideal Views* (1855).

47. Zimmermann, *Die Wunder der Urwelt*, 7th ed. (1855) p. 2, "Die Archive der Vorwelt." I have not seen the earlier editions of this work.

48. Duncan, *Pre-Adamite Man* (1860), pp. 193–94. The remainder of this long "Explanation" describes the extinct animals in detail, citing the standard scientific sources.

49. W. Hawkins, "Visual Education" (1854), pp. 445–46.

50. *Illustrated London News*, vol. 24, p. 22, issue of 7 January 1854.

51. *Illustrated London News*, vol. 23, p. 599, issue of 31 December 1853.

52. W. Hawkins, "Visual Education" (1854), p. 445.

53. Buckland, *Geology and Mineralogy*, new ed. (1858), pp. 33, 35.

54. Unger, *Urwelt*, 2d ed. (1858), preface.

55. Unger, *Urwelt*, 2d ed. (1858), explanation of Taf. A.

56. Unger, *Urwelt*, 2d ed. (1858), explanation of Taf. B.

57. Tennant, "Waterhouse Hawkins's Restorations" (July 1860).

58. Boitard, *Paris avant les hommes* (1861), p. 65.

59. Boitard, *Paris avant les hommes* (1861), pp. 245–47.

60. Figuier, *La terre avant le Déluge* (1863), p. 40; trans. from *World before the Deluge* (1865), p. 40.

61. Figuier, *La terre avant le Déluge* (1863), p. 52; the text of the English edition at this point is by its editor, not Figuier.

62. Figuier, *La terre avant le Déluge* (1863), p. 63; trans. from *World before the Deluge* (1867), pp. 104–5.

63. Figuier, *La terre avant le Déluge* (1863), p. 80.

64. Figuier, *La terre avant le Déluge*, 4th ed. (1865), pp. 93–94; trans. from *World before the Deluge* (1865), pp. 131–32.

65. Figuier, *La terre avant le Déluge*, 4th ed. (1865), pp. 101–02; trans. from *World before the Deluge* (1865), pp. 138–40.

66. Figuier, *La terre avant le Déluge* (1863), p. 105; trans. from *World before the Deluge* (1865), pp. 151–52.

67. Figuier, *La terre avant le Déluge* (1863), pp. 123–24; trans. from *World before the Deluge* (1865), pp. 172–73.

68. Figuier, *La terre avant le Déluge* (1863), pp. 133–34; the editor of the English translation introduces much material of his own at this point (*World before the Deluge*, pp. 175–76).

69. Figuier, *La terre avant le Déluge* (1863), p. 154; trans. from *World before the Deluge* (1865), p. 203.

70. Figuier, *La terre avant le Déluge* (1863), p. 162.

71. Figuier, *La terre avant le Déluge* (1863), p. 184; trans. from *World before the Deluge* (1865), p. 221–22.

72. Figuier, *La terre avant le Déluge*, 4th ed. (1865), pp. 194–97; trans. from *World before the Deluge* (1865), pp. 223–24.

73. Figuier, *La terre avant le Déluge*, 4th ed. (1865), pp. 205–6; trans. from *World before the Deluge* (1865), p. 230.

74. Figuier, *La terre avant le Déluge* (1863), p. 201; trans. from *World before the Deluge* (1865), pp. 258–59.

75. Figuier, *La terre avant le Déluge* (1863), p. 225; trans. from *World before the Deluge* (1865), p. 270.

76. Figuier, *La terre avant le Déluge* (1863), pp. 248–49; trans. from *World before the Deluge* (1865), p. 290.

77. Figuier, *La terre avant le Déluge* (1863), p. 268; trans. from *World before the Deluge* (1865), p. 311.

78. Figuier, *La terre avant le Déluge* (1863), p. 290; trans. from *World before the Deluge* (1865), p. 333.

79. Figuier, *La terre avant le Déluge*, 4th ed. (1865), p. 360. This text was adapted from the first edition (p. 283), simply by substituting "Quaternary" for "Pliocene" and making other, minor changes.

80. Figuier, *La terre avant le Déluge* (1863), p. 322; trans. from *World before the Deluge* (1865), pp. 368–69.

81. Figuier, *La terre avant le Déluge* (1863), pp. 326–28; trans. from *World before the Deluge* (1865), pp. 375–76.

SOURCES FOR FIGURES AND TEXTS

82. Figuier, *La terre avant le Déluge* (1863), pp. 358–62; trans. from *World before the Deluge* (1865), pp. 415–16, 419–20.

83. Figuier, *La terre avant le Déluge*, 4th ed. (1865), p. 399.

84. Figuier, *La terre avant le Déluge* (1863), pp. 365–71; trans. from *World before the Deluge* (1865), pp. 429–32.

85. Fraas, *Sündfluth!* (1866), "Prospectus." It is immaterial in the present context whether this was written by Fraas or his publisher; in either case Fraas must have approved it.

86. Figuier, *La terre avant le Déluge* (1863); trans. from *World before the Deluge* (1865), pp. 1–2.

Bibliography

Primary Sources

Agassiz, Louis. 1840. *Études sur les glaciers*. Neuchâtel: The author.

————. 1967. *Studies on glaciers, preceded by the Discourse of Neuchâtel*, ed. Albert V. Carozzi. New York and London: Hafner.

Ansted, David T., [James] Tennant, and Walter Mitchell. 1855. *Geology, mineralogy and crystallography: Being a theoretical, practical and descriptive view of inorganic nature. The form and classification of crystals, and a chemical arrangement of minerals*. London: Houlston and Stoneman.

Bible. 1836. *Allgemeine, wohlfeile Bilder-Bibel für die Katholiken, oder die ganze heilige Schrift das alten und neuen Testaments . . . mit mehr als 500 schonen in den Texte eingedruckten Abbildungen. . . .* 2 vols. Leipzig: Baumgartner.

Bible. 1836–38. *The pictorial Bible; being the Old and New Testaments according to the Authorized Version: Illustrated with many hundred woodcuts. . . .* 3 vols. London: C. Knight.

Boblaye, [Emile Le Puillon de]. 1834. Animaux fossiles; Animaux perdus. *Dictionnaire pittoresque d'histoire naturelle* 1: 2–5, pl. 24. Paris: Bureau de Souscription.

Boitard, [Pierre]. 1861. *Études antediluviennes. Paris avant les hommes, l'homme fossile, etc., histoire naturelle du globe terrestre. Illustrée d'après les dessins de l'auteur*. Paris: Passard.

Brewer, [Ebenezer Cobham]. 1860. *Theology in science; containing the following subjects: geology, physical geography . . . and shewing the wisdom and goodness of God in their respective phenomena. For the use of schools and of private readers*. London: Jarrold and Sons.

Bru, Juan Bautista. 1796. Descripción del esqueleto en particular, según las observaciones hechas al tiempo de armarle y colocarle en este Real Gabinete. In *Descripción del esqueleto de un quadrúpedo muy porpulento y rara, que se conserva en el Real Gabinete de Historia Natural de Madrid*, ed. José Garriga, pp. 1–16, pls. 1–5. Madrid: Viuda de Ibarra.

Buckland, Francis Trevelyan. 1857. *Curiosities of natural history*. London: Richard Bentley.

————. 1860. *Curiosities of natural history*. 2d series. London: Richard Bentley.

Buckland, William. 1822. Account of an assemblage of fossil teeth and bones of elephant, rhinoceros, hippopotamus, bear, tiger, and hyaena, and sixteen other animals, discovered in a cave at Kirkdale, Yorkshire, in the year 1821. *Philosophical Transactions of the Royal Society of London* 1822: 171–236, pls. 15–26.

————. 1823. *Reliquiae Diluvianae; or, observations on the organic remains contained in caves, fissures, and diluvial gravel, and on other geological phenomena, attesting to the action of an universal deluge*. London: John Murray.

————. 1824. Notice on the Megalosaurus or great fossil lizard of Stonesfield. *Transactions of the Geological Society of London*, 2d series, 1 (2): 390–96, pls. 40–44.

————. 1835a. On the discovery of a new species of Pterodactyle in the Lias at Lyme Regis. *Transactions of the Geological Society of London*, 2d series, 3 (1): 217–22, pl. 27 (read 6 February 1829).

———. 1835b. On the discovery of coprolites, or fossil faeces, in the Lias at Lyme Regis, and in other Formations. *Transactions of the Geological Society of London*, 2d series, 3 (1): 223–36, pls. 28–31 (read 6 February 1829).

———. 1836. *Geology and mineralogy considered with reference to natural theology.* 2 vols. London: William Pickering.

———. 1837. *Geology and mineralogy. . . .* 2d ed. London: William Pickering.

———. 1858. *Geology and mineralogy. . . .* "New" ed. [by Francis T. Buckland]. 2 vols. London: George Routledge.

Buffon, George Leclerc, [Comte] de. 1778. *Les époques de la nature.* Supplément 5 of *Histoire naturelle*, 254 pp.

Conybeare, William Daniel. 1824. On the discovery of an almost perfect skeleton of the Plesiosaurus. *Transactions of the Geological Society of London*, 2d series, 1 (2): 381–89, pls. 48–49.

Conybeare, William Daniel, and Henry Thomas De la Beche. 1821. Notice of a discovery of a new fossil animal, forming a link between the ichthyosaurus and the crocodile; together with general remarks on the osteology of the ichthyosaurus. *Transactions of the Geological Society of London* 1: 558–94, pls. 40–42.

Conybeare, William Daniel, and William Phillips. 1822. *Outlines of the geology of England and Wales, with an introductory compendium of the general principles of that science, and comparative views of the structure of foreign countries.* Part 1 [all issued]. London: William Phillips.

Cuvier, Georges. 1796. Notice sur le squelette d'une très-grande espèce de quadrupède inconnue jusqu'à présent, trouvé au Paraguay, et déposé au cabinet d'histoire naturelle de Madrid, redigée par G. Cuvier. *Magasin encyclopédique*, 2e année, 1: 303–10, 2 pls.

———. 1804–8. Sur les espèces d'animaux dont proviennent les os fossiles répandus dans la pierre à plâtre des environs de Paris. *Annales du Muséum d'Histoire Naturelle* 3: 275–303, 364–87, 442–72; 4: 66–75; 6: 253–83; 9: 10–44, 89–102, 205–15, 272–82; 12: 271–84.

———. 1806. Sur le grande Mastodonte, animal très-voisin de l'elephant, mais à mâchelières hérissées de gros tubercles, dont on trouve les os en divers endroits des deux continens, et surtout près des bords de l'Ohio, dans l'Amérique Septentrionale, improprement nommé Mammouth par les Anglais et par les habitans des États-Unis. *Annales du Muséum d'Histoire Naturelle* 8: 270–312, 7 pls.

———. 1812. *Recherches sur les ossemens fossiles de quadrupèdes, où l'on rétablit les caractères de plusieurs espèces d'animaux que les révolutions du globe paroissent avoir détruites.* 4 vols. Paris: Déterville.

———. 1821–24. *Recherches sur les ossemens fossiles, où l'on rétablit les caractères de plusieurs espèces d'animaux dont les révolutions du globe ont détruites les espèces. Nouvelle édition, entièrement refondue, et considérablement augmentée.* 5 vols. in 7. Paris: Dufour and d'Ocagne.

Darwin, Charles. 1859. *On the origin of species by means of natural selection, or the preservation of favoured races in the struggle for life.* London: John Murray.

Daubeny, Charles G. B. 1869. *Fugitive poems connected with natural history and physical science. Collected by the late C. G. B. Daubeny.* Oxford and London: James Parker.

De la Beche, Henry T. 1831. *A geological manual.* London: Treuttel and Würtz, Treuttel Jun. and Richter.

———. 1832a. *A geological manual.* 2d ed. London: Treuttel and Würtz, Treuttel Jun. and Richter.

———. 1832b. *Handbuch der Geognosie.* [Translated from second edition by Heinrich von Dechen.] Berlin: Duncker & Humblot.

———. 1833. *Manuel géologique.* [Translated from second edition by A. J. M. Brochant de Villiers.] Paris: F. G. Levrault.

Doré, Gustave. 1866a. *La sainte Bible selon la Vulgate. Traduction nouvelle, avec les dessins de G. Doré.* 2 vols. Tours.

———. 1866b. *The Holy Bible containing the Old and New Testaments, according to the Authorized Version. With illustrations by Gustave Doré.* London and New York: Cassell, Petter and Galpin.

Duncan, Isabella. 1860. *Pre-Adamite man; or, the story of our old planet & its inhabitants, told by scripture and science.* London: Saunders, Otley.

Faujas de Saint-Fond, Barthélemy. [1798–99.] *Histoire naturelle de la Montagne de Saint-Pierre de Maestricht.* Paris: H. J. Jansen.

Figuier, Louis. 1851. *Exposition et histoire des principales découvertes scientifiques modernes.* Paris: Masson.

———. 1860. *Histoire des merveilleux dans les temps modernes.* 4 vols. Paris: Hachette.

———. 1863. *La terre avant le Déluge: Ouvrage contenant 24 vues idéales de paysages de l'ancien monde dessinées par Riou.* Paris: Hachette.

———. 1865a. *La terre avant le Déluge. . . .* 4th ed. Paris: Hachette.

———. 1865b. *The world before the Deluge, containing twenty-five ideal landscapes of the ancient world, designed by [Edouard]*

Riou . . . translated from the fourth French edition. Edited by W. S. O[rmerod]. London: Chapman and Hall.

———. 1867a. *La terre avant le Déluge*. . . . 6th ed.

———. 1867b. *The world before the Deluge: A new edition, the geological portion carefully revised, and much original matter added, by Henry W. Bristow, F.R.S.* . . . London: Chapman and Hall.

Flammarion, Camille. 1886. *Le monde avant la création de l'homme. Origines de la terre. Origines de la vie. Origines de l'humanité.* Paris: C. Marpon and E. Flammarion.

Fraas, Oskar von. 1866. *Vor der Sündfluth! Eine Geschichte der Urwelt. Mit vielen Abbildungen ausgestorbener Thiergeschlechter und urweltlicher Landschaftsbilder.* Stuttgart: Carl Hoffmann.

Goldfuss, August. 1826–44. *Petrefacta Germaniae . . . Abbildungen und Beschreibungen der petrefacten Deutschlands und der angränzenden Länder unter Mitwerkung des Herrn Grafen Georg zu Münster, herausgegeben von August Goldfuss.* 3 vols. Dusseldorf: Arnz and Comp.

———. 1831. Beiträge zur Kenntnis verschiedener Reptilien der Vorzeit. *Nova acta physico-medica Academiae Caesareae Leopoldino-Carolinae* 15 (1): 61–128, 7 pls.

Goldsmith, Oliver. 1832. *A history of the earth and animated nature . . . to which is subjoined an appendix, by Captn. Thomas Brown.* . . . 4 vols. Glasgow: Archibald Fullerton.

Goodrich, Samuel Griswold. 1859. *Illustrated natural history of the animal kingdom, being a systematic and popular description of the habits, structure, and classification of animals from the highest to the lowest forms, with their relations to agriculture, commerce, manufactures and the arts.* New York: Derby and Jackson. [See also Parley, Peter.]

Guérin[-Ménéville], Félix Edouard, ed. 1834–39. *Dictionnaire pittoresque d'histoire naturelle.* 9 vols. Paris: Bureau de Souscription.

Hartmann, Carl. 1841. *Die Schöpfungswunder der Unterwelt. Interessante Schilderungen der berühmsten Höhlen, Quellen, Erdbeben, Vulkane, Bergwerke, Versteinerungen und andere Merkwürdigkeiten. Für Jung und Alt.* 2 vols. Stuttgart: J. Schieble.

Hawkins, B[enjamin] Waterhouse. 1854. On visual education as applied to geology. *Journal of the Society of Arts* 2: 444–49.

Hawkins, Thomas. 1834. *Memoirs of ichthyosauri and plesiosauri, extinct monsters of the ancient earth.* London: Rolfe and Fletcher.

———. 1840. *The book of the great sea-dragons, ichthyosauri and plesiosauri, gedolim tanimim, of Moses. Extinct monsters of the ancient earth.* London: William Pickering.

———. 1844. *The wars of Jehovah in Heaven, Earth and Hell.* London: Francis Baisler.

Heck, Johann Georg. 1849. *Bilder-Atlas zum Conversations-Lexicon. Ikonographische Encyclopädie der Wissenschaft und Kunste. Entworfen und nach den vorzüglichsten Quellen bearbeitet von Johann Georg Heck.* Leipzig: F. U. Brockhaus.

Hutton, James. 1795. *Theory of the earth, with proofs and illustrations.* 2 vols. Edinburgh: William Creech.

Kaup, Johann Jacob. 1832–39. *Description d'ossements fossiles de mammifères inconnus jusqu'à présent, qui se trouvent au Musée grand-ducal de Darmstadt; avec figures lithographiées.* Darmstadt: J. P. Diehl.

Kircher, Athanasius. 1675. *Arca Noë in tres libros digesta, quorum I De rebus ante Diluvium, II De iis, quae ipso diluvio, eiusque duratione, et III De iis, quae post diluvium a Noemo gesta sunt. Quae omnia nova Methodo, nec non Argumentorum varietate, explicantur, & demonstrantur.* Amsterdam: Johann Jansson.

Klipstein, August von, and Johann Jacob Kaup. 1836. *Beschreibung und Abbildung von dem in Rheinhessen aufgefundenen colossalen Schädel des Dinotherii gigantei, mit geognostischen Mittheilungen über die knochenführenden Bildungen des mittelrheinischen Tertiärbeckens.* Darmstadt: Johann Philip.

[La Peyrère, Isaac.] 1655. *Prae-Adamitae.* n.p.

Lyell, Charles. 1830–33. *Principles of geology, being an attempt to explain the former changes of the earth's surface, by reference to causes now in operation.* 3 vols. London: John Murray.

———. 1863. *The geological evidences of the antiquity of man. With remarks on theories of the origin of species by variation.* London: John Murray.

Mantell, Gideon Algernon. 1831. The geological age of reptiles. *Edinburgh New Philosophical Journal* 11: 181–85.

———. 1838. *The wonders of geology; or, a familiar exposition of geological phenomena. Being the substance of a course of lectures delivered at Brighton, from notes taken by G. F. Richardson, Curator of the Mantellian Museum etc.* 2 vols. London: Relfe and Fletcher.

———. 1839. *Die Phänomene der Geologie leichtfasslich in Vorlesungen entwickelt . . . Deutsch herausgegeben von Dr Joseph Burkart.* 2 vols. Bonn: Henry and Cohen.

———. 1864. *The wonders of geology.* . . . 8th ed. London.

Martin, John. 1828. *A descriptive catalogue of the engraving of the Deluge.* London: Plummer and Brewer.

———. 1838. *Illustrations of the Bible. Designed and engraved by John Martin.* London: Charles Tilt.

Milner, Thomas. 1846. *The gallery of nature, a pictorial and descriptive tour through Creation, illustrative of the wonders of astronomy, physical geography and geology.* London.

Murchison, Roderick Impey. 1839. *The Silurian system, founded on geological researches in the countries of Salop, Hereford [etc.]; with descriptions of the coal-fields and overlying formations.* London: John Murray.

Orbigny, Alcide d'. 1849–52. *Cours élémentaire de paléontologie et de géologie stratigraphique.* 2 vols. Paris: Victor Masson.

Owen, Richard. 1841. On the teeth of species of the genus *Labyrinthodon* (*Mastodonsaurus* of Jaeger) from the German Keuper formation and the Lower Sandstone of Warwick and Leamington. *Transactions of the Geological Society of London*, 2d series, 6 (2): 503–13.

———. 1842. Report on British fossil reptiles. Part 2. *Reports of the British Association for the Advancement of Science* 1841: 60–204.

———. 1854. *Geology and inhabitants of the ancient world.* Crystal Palace Guidebooks. London: Crystal Palace Library.

Parkinson, James. 1804–11. *Organic remains of a former world. An examination of the mineralized remains of the vegetables and animals of the antediluvian world; generally termed extraneous fossils.* 3 vols. London: Sherwood, Neely and Jones.

Parley, Peter [Samuel Griswold Goodrich]. [1837.] *Peter Parley's Wonders of earth sea and sky.* London: Darton and Hodge.

[Phillips, John.] 1833. Organic remains restored. *Penny Magazine* 2 (100): 409–10.

Pictet, François-Jules. 1844–46. *Traité élémentaire de paléontologie ou histoire naturelle des animaux fossiles considerées dans leurs rapports zoologiques et géologiques.* 4 vols. Geneva: Jules-Guillaume Fick.

Reyer, Alexander. 1871. *Leben und Wirken des Naturhistorikers Dr Franz Unger, Professor der Pflanzen-Anatomie und Physiologie.* Graz: Leuschner and Lubensky.

Reynolds, James. 1849. The ante-diluvian world [and] Popular geology. [Two broadsheets.] London: James Reynolds.

Richardson, George Fleming. 1838. *Sketches in prose and verse (second series), containing visits to the Mantellian Museum, descriptive of that collection: Essays, tales, poems, &c. &c.* London: Rolfe and Fletcher.

———. 1842. *Geology for beginners, comprising a familiar explanation of geology, and its associate sciences, mineralogy, physical geology, fossil conchology, fossil botany, and palaeon-tology. Including directions for forming collections and generally cultivating the science, with a succinct account of the several geological formations.* London: Hippolyte Baillière.

———. 1843. *Geology for beginners.* . . . 2d ed. London: Longman, Brown, Green and Longmans.

Scheuchzer, Johann Jakob. 1709. *Herbarium Diluvianum collectum a Johanne Jacobo Scheuchzero.* . . . Zurich: David Gessner.

———. 1726. *Homo Diluvii testis.* Zurich.

———. 1731–35a. *Physica sacra Johannis Jacobi Scheuchzeri . . . iconibus aeneis illustrata procurante & sumtus suppeditante Johanne Andrea Pfeffel.* . . . Augsburg and Ulm.

———. 1731–35b. *Kupfer-Bibel, in welche die Physica Sacra, oder geheiligte Natur-Wissenschaft derer in Heil. Schrifft vorkommenden natürlichen Sachen, deutlich erklart und bewahrt von J. J. Scheuchzer.* . . . Augsburg.

———. 1732–37. *Physique sacrée, ou histoire naturelle de la bible. Traduit du latin de J. J. Scheuchzer, enrichie de figures en taille-douce, gravée par les soins de Jean-André Pfeffel, graveur de S. M. Imperiale.* Amsterdam: P. Schenk and P. Mortier.

———. 1735–39. *Geestelijke natuurkunde, uitgegeven in de Latijnsche taal door Johann Jakob Scheuchzer . . . in't Nederduitsch vertaalt door F. H. J. van Halen.* Amsterdam: P. Schenk.

Schroeter, Johann Samuel. 1774–76. *Beyträge zur Naturgeschichte sonderlich des Mineralreichs, aus ungedrukten Briefen gelehrtes Naturforscher und aufmerksamer Freunde der Natur.* Altenberg: Richter.

Tennant, James. 1860. *Key to a coloured lithographic plate of Waterhouse Hawkins's restorations of extinct animals.* London: Tennant.

Trimmer, Joshua. 1841. *Practical geology and mineralogy; with instructions for the qualitative analysis of minerals.* London: John W. Parker.

Unger, Franz-Xaver. 1841–47. *Chloris protogaea. Beiträge zur Flora der Vorwelt.* Leipzig: Engelmann.

———. [1851.] *Die Urwelt in ihren verschiedenen Bildungsperioden. 14 landschaftliche Darstellungen mit erlauternden Text. Le monde primitif à ses differentes époques de formation. 14 paysages avec texte explicatif.* Vienna: Beck.

———. [1855.] *Ideal views of the primitive world, in its geological and palaeontological phases, illustrated by fourteen photographic plates, being an introduction to the series. Photography in its application to palaeontology.* (Highley's Library of Science and Art. Section 2: Natural History.) London: Samuel Highley.

―――. 1858. *Die Urwelt . . . Le monde primitif. . . .* 2d ed. Leipzig: T. O. Weigel.

―――. [1864.] *Ideal views. . . .* 2d ed. London: Samuel Highley.

Verne, Jules. [1865.] *Cinq semaines en ballon. Voyage de découvertes en Afrique par trois Anglais.* Paris: J. Hetzel.

Zimmermann, W. F. A. [W. F. Volliner]. 1855–58. *Populaires Handbuch der Physikalischen Geographie.* 5th ed. 3 vols in 4. Berlin: Gustav Hempel.

―――. 1855. *Die Wunder der Urwelt. Eine populäre Darstellung der Geschichte der Schöpfung und des Urzustandes unsere Weltkorpers so wie der verschiedenen Entwicklungs-Perioden seine Oberfläche, seine Vegetation und seiner Bewohner bis auf die Jetztzeit. Nach den Resultäten der Forschung und Wissenschaft bearbeitet.* 7th ed. [Vol. 3, part 1 of *Physicalische Geographie.*] Berlin: Gustav Hempel.

―――. 1857. *Le Monde avant la création de l'homme, ou le berceau de l'univers. Histoire populaire de la création et des transformations du globe, racontée aux gens du monde.* Paris: Schultz and Thuillié.

―――. 1869. *The wonders of the primitive world. A description of the history of creation; and of the original state of our planet. . . .* The People's Library. New York: Charles Pfirshing.

Secondary Sources

Abel, Othenio. 1922. *Lebensbilder aus der Tierwelt der Vorzeit.* Jena: Fischer.

―――. 1925. *Geschichte und Methode der Rekonstruktion vorzeitliche Wirbeltiere.* Jena: Fischer.

Alfrey, Nicholas. 1990. Landscape and the Ordnance Survey, 1795–1820. In *Mapping the landscape: Essays on art and cartography,* ed. Nicholas Alfrey and Stephen Daniels, pp. 23–27, pls. 19–24. Nottingham: University Art Gallery.

Allen, David. 1976. *The naturalist in Britain: A social history.* London: Allen Lane.

Allen, Don Cameron. 1949. *The legend of Noah: Renaissance rationalism in art, science, and letters.* Urbana: University of Illinois Press.

Alpers, Svetlana. 1983. *The art of describing: Dutch art in the seventeenth century.* Chicago and London: University of Chicago Press and John Murray.

Altick, Richard D. 1978. *The shows of London.* Cambridge, Mass.: Harvard University Press.

Bann, Stephen. 1984. *The clothing of Clio: A study of the representation of history in nineteenth-century Britain and France.* Cambridge: Cambridge University Press.

Barber, Lynn. 1980. *The heyday of natural history, 1820–1870.* Garden City, N.J.: Doubleday.

Baxandall, Michael. 1972. *Painting and experience in fifteenth-century Italy: A primer in the social history of pictorial style.* Oxford: Oxford University Press.

―――. 1985. *Patterns of intention: On the historical explanation of pictures.* New Haven: Yale University Press.

Bermingham, Ann. 1986. *Landscape and ideology: The English rustic tradition, 1740–1860.* Berkeley: University of California Press.

Blunt, Wilfrid. 1976. *The Ark in the Park: The Zoo in the nineteenth century.* London: Hamish Hamilton.

Bowler, Peter J. 1976. *Fossils and progress: Paleontology and the idea of progressive evolution in the nineteenth century.* New York: Science History.

―――. 1988. *The non-Darwinian revolution: Reinterpreting a historical myth.* Baltimore: Johns Hopkins University Press.

Browne, Janet. 1983. *The secular Ark: Studies in the history of biogeography.* New Haven: Yale University Press.

Bryson, Norman. 1981. *Word and image: French painting of the ancien regime.* Cambridge: Cambridge University Press.

Cardot, Fabienne, ed. 1985. *Louis Figuier: Les merveilles de l'électricité. Textes choisis.* Paris: Association pour l'Histoire de l'Électricité en France.

Cohen, C., and J. J. Hublin. 1989. *Boucher de Perthes, 1788–1868: Les origines romantiques de la préhistoire.* Paris: Belin.

Coleman, William. 1963. *Georges Cuvier, zoologist: A study in the history of evolution theory.* Cambridge: Harvard University Press.

Corsi, Pietro. 1978. The importance of French transformist ideas for the second volume of Lyell's *Principles of geology. British Journal for the History of Science* 11: 221–44.

―――. 1988. *The age of Lamarck: Evolutionary theories in France, 1790–1830.* Berkeley: University of California Press.

Curwen, E. Cecil, ed. 1940. *The journal of Gideon Mantell, surgeon and geologist, covering the years 1818–1852.* London: Oxford University Press.

Czerkas, Sylvia J., and Everett C. Olson, eds. 1987. *Dinosaurs past and present.* 2 vols. Seattle and London: University of Washington Press and Natural History Museum of Los Angeles County.

Dean, Dennis R. 1990. A bicentenary retrospective on Gideon Algernon Mantell (1790–1852). *Journal of Geological Education* 38: 434–43.

Delair, Justin B., and William A. S. Sarjeant. 1975. The earliest discoveries of dinosaurs. *Isis* 66: 5–25.

Desmond, Adrian J. 1974. Central Park's fragile dinosaurs. *Natural History* 83: 64–71.

———. 1979. Designing the dinosaur: Richard Owen's response to Robert Edmond Grant. *Isis* 70: 224–34.

———. 1989. *The politics of evolution: Morphology, medicine, and reform in radical London.* Chicago: University of Chicago Press.

Feaver, William. 1975. *The art of John Martin.* Oxford: Clarendon Press.

Fischer, Hans. 1973. *Johann Jakob Scheuchzer: Naturforscher und Arzt.* Zurich: Leeman.

Fyfe, Gordon, and John Law, eds. 1988. *Picturing power: Visual depictions and social relations.* London: Routledge.

Godwin, Joscelyn. 1979. *Athanasius Kircher: A Renaissance man and the quest for lost knowledge.* London: Thames and Hudson.

Gohau, Gabriel. 1990. *Les sciences de la terre aux XVIIe et XVIIIe siècles: Naissance de la géologie.* Paris: Albin Michel.

———. 1991. *History of geology.* New Brunswick, N.J.: Rutgers University Press. [Translation of *Histoire de la géologie.* Paris: Éditions La Découverte, 1987.]

Gould, Stephen J. 1987. *Time's arrow, time's cycle: Myth and metaphor in the discovery of geological time.* Cambridge: Harvard University Press.

———. 1989. *Wonderful life: The Burgess Shale and the nature of history.* New York: W. W. Norton.

Grafton, Anthony T. 1991. *Defenders of the text: The traditions of scholarship in an age of science, 1450–1800.* Cambridge: Harvard University Press.

Grayson, Donald K. 1983. *The establishment of human antiquity.* New York: Academic Press.

Hacking, Ian. 1983. *Representing and intervening: Introductory topics in the philosophy of natural science.* Cambridge: Cambridge University Press.

Haraway, Donna. 1989. *Primate visions: Gender, race, and nature in the world of modern science.* New York: Routledge.

Haubold, Hartmut, and Oskar Kuhn. 1977. *Lebensbilder und Evolution fossiler Saurier: Amphibien und Reptilien.* Wittenberg: A. Ziemsen.

Herrmann, Luke. 1973. *British landscape painting in the eighteenth century.* London: Faber.

Howe, S. R., T. Sharpe, and H. S. Torrens. 1981. *Ichthyosaurs: A history of fossil 'sea-dragons.'* Cardiff: National Museum of Wales.

Inkster, Ian. 1979. London science and the Seditious Meetings Act of 1817. *British Journal for the History of Science* 12: 192–96.

Ivins, William M., Jr. 1953. *Prints and visual communication.* London: Routledge and Kegan Paul.

Jahn, Melvyn. 1969. Some notes on Dr Scheuchzer and *Homo Diluvii testis.* In *Toward a history of geology,* ed. Cecil J. Schneer, pp. 192–213. Cambridge: M.I.T. Press.

Johnstone, Christopher. 1974. *John Martin.* London: Academy Editions.

Jordanova, Ludmilla. 1989. Objects of knowledge: A historical perspective on museums. In *The new museology,* ed. Peter Vergo, pp. 22–40. London: Reaktion.

Jussim, Estelle. 1974. *Visual communication and the graphic arts: Photographic technologies in the nineteenth century.* New York: R. R. Bowker.

Kirchheimer, Franz. 1982. Die Einführung des Naturselbstdruckes und der Photographie in die erdwissenschaftliche Dokumentation. *Zeitschrift der Deutsche Geologische Gesellschaft* 133: 1–117.

Koenigswald, Wighart von. 1982. Das Dinotherium von Eppelsheim. *Alzeyer Geschichtblätter* Sonderheft 8: 17–29.

Langer, Wolfhart. 1971. Georg August Goldfuss. Ein biographischer Beitrag. *Bonner Geschichtsblätter* 23: 229–43.

———. 1990. Frühe Bilder aus der Vorzeit. *Fossilien* 5: 202–5.

Latour, Bruno. 1986. Visualisation and cognition: Thinking with eyes and hands. *Knowledge and Society* 6: 1–40.

Latour, Bruno, and J. de Noblet, eds. 1983. *Les "vues" de l'esprit.* Paris. [*Culture technique* 14.]

Laudan, Rachel. 1977. Ideas and organizations in British geology: A case study in institutional history. *Isis* 68: 527–38.

Laurent, Goulven. 1987. *Paléontologie et évolution en France de 1800 à 1860: Une histoire des idées de Cuvier et Lamarck à Darwin.* Paris: Éditions du C.T.H.S.

———. 1989. Idées sur l'origine de l'homme en France de 1800 à 1871 entre Lamarck et Darwin. *Bulletin et mémoires de la Société d'Anthropologie de Paris,* new ser., 1: 105–30.

Livingstone, David N. 1986. Preadamites: The history of an idea from heresy to orthodoxy. *Scottish Journal of Theology* 40: 41–66.

Lopez Piñero, José M. 1988. Juan Bautista Bru (1740–1799)

and the description of the genus *Megatherium. Journal of the History of Biology* 21: 146–63.

Lynch, Michael. 1985. Discipline and the material form of images: An analysis of scientific visibility. *Social Studies of Science* 15: 37–66.

Lynch, Michael, and Steve Woolgar, eds. 1988. *Representation in scientific practice.* Dordrecht: Kluwer Academic Press. [*Human Studies* 11 (2/3).]

McCartney, Paul J. 1977. *Henry De la Beche: Observations on an observer.* Cardiff: Friends of the National Museum of Wales.

McKee, Alexander. 1968. *History under the sea.* London: Hutchinson.

McPhee, John. 1981. *Basin and range.* New York: Farrar, Strauss, Giroux.

Maré, Eric de. 1980. *Victorian wood-block illustrators.* London: Gordon Fraser.

Merrill, Lynn L. 1989. *The romance of Victorian natural history.* New York: Oxford University Press.

Miller, David Philip. 1986. Method and the "micropolitics" of science: The early years of the Geological and Astronomical Societies of London. In *The politics and rhetoric of scientific method,* ed. J. A. Schuster and R. R. Yeo, pp. 227–57. Dordrecht: Reidel.

Morris, A. D. 1989. *James Parkinson: His life and times.* Boston: Birkhäuser.

Nissen, Claus. 1964–. *Zoologische Buchillustration: Ihre Bibliographie und Geschichte.* Stuttgart: Anton Hiersemann.

Outram, Dorinda. 1984. *Georges Cuvier: Vocation, science and authority in post-Revolutionary France.* Manchester: Manchester University Press.

Padian, Kevin. 1987. The case of the bat-winged pterosaur: Typological taxonomy and the influence of pictorial representation on scientific perception. In *Dinosaurs past and present,* ed. Sylvia J. Czerkas and Everett C. Olson, vol. 2 [unpaginated].

Popkin, Richard H. 1987. *Isaac La Peyrère (1596–1676): His life, work and influence.* Leiden: Brill.

Prest, John. 1981. *The Garden of Eden: The botanic garden and the recreation of Paradise.* New Haven: Yale University Press.

Rappaport, Rhoda. 1978. Geology and orthodoxy: The case of Noah's Flood in eighteenth-century thought. *British Journal for the History of Science* 11: 1–18.

———. 1982. Borrowed words: Problems of vocabulary in eighteenth-century geology. *British Journal for the History of Science* 15: 27–44.

Rehbock, Philip F. 1980. The Victorian aquarium in ecological and social perspective. In *Oceanography: The past,* ed. M. Sears and D. Merriman, pp. 522–39. New York: Springer-Verlag.

Ritvo, Harriet. 1985. Learning from animals: Natural history for children in the eighteenth and nineteenth centuries. *Children's Literature* 13: 72–93.

———. 1987. *The animal estate: The English and other creatures in the Victorian age.* Cambridge: Harvard University Press.

Roger, Jacques. 1962. Buffon: Les Époques de la Nature. Édition critique. *Mémoires du Muséum Nationale d'Histoire Naturelle,* sér. C. 10.

———. 1989. *Buffon: Un philosophe au Jardin du Roi.* Paris: Fayard.

Roselle, Daniel. 1968. *Samuel Griswold Goodrich, creator of Peter Parley. A study of his life and work.* Albany: S.U.N.Y. Press.

Rosen, Charles, and Henri Zerner. 1984. *Romanticism and realism: The mythology of nineteenth-century art.* London: Faber and Faber.

Rossi, Paolo. 1984. *The dark abyss of time: The history of the earth and the history of nations from Hooke to Vico.* Chicago: University of Chicago Press.

Rudwick, Martin J. S. 1963. The foundation of the Geological Society of London: Its scheme for cooperative research and its struggle for independence. *British Journal for the History of Science* 1: 325–55.

———. 1971. Uniformity and progression: Reflections on the structure of geological theory in the age of Lyell. In *Perspectives in the history of science and technology,* ed. Duane H. D. Roller, pp. 209–27. Norman: Oklahoma University Press.

———. 1972. *The meaning of fossils: Episodes in the history of palaeontology.* London and New York: MacDonald and American Elsevier.

———. 1975. Caricature as a source for the history of science: De la Beche's anti-Lyellian sketches of 1831. *Isis* 66: 534–60.

———. 1976. The emergence of a visual language for geological science, 1760–1840. *History of Science* 14: 149–95.

———. 1985. *The great Devonian controversy: The shaping of scientific knowledge among gentlemanly specialists.* Chicago: University of Chicago Press.

———. 1989. Encounters with Adam, or at least the hyaenas: Nineteenth-century visual representations of the deep past. In *History, humanity and evolution: Essays for John C.*

Greene, ed. James R. Moore, pp. 231–51. Cambridge: Cambridge University Press.

Rupke, Nicolaas A. 1983. *The great chain of history: William Buckland and the English school of geology (1814–1849)*. Oxford: Clarendon Press.

Schivelbusch, Wolfgang. 1989. *Railway journey: The industrialization of time and place in the nineteenth century*. Berkeley: University of California Press.

Secord, James A. 1986a. *Controversy in Victorian geology: The Cambrian-Silurian dispute*. Princeton: Princeton University Press.

———. 1986b. The Geological Survey of Great Britain as a research school, 1839–1855. *History of Science* 24: 223–75.

Shapin, Steven. 1989. Science and the public. In *A companion to the history of modern science*, ed. R. C. Olby, G. N. Cantor, J. R. R. Christie, and M. J. S. Hodge, pp. 990–1007. London: Routledge.

———. 1989. The invisible technician. *American scientist* 77: 554–63.

Shapin, Steven, and Simon Schaffer. 1985. *Leviathan and the air pump*. Princeton: Princeton University Press.

Sorensen, Colin. 1989. Theme parks and time machines. In *The new museology*, ed. Peter Vergo, pp. 60–73. London: Reaktion.

Stafford, Barbara Maria. 1984. *Voyage into substance: Art, science, nature, and the illustrated travel account, 1760–1840*. Cambridge: M.I.T. Press.

Thackray, John C. 1976. James Parkinson's *Organic remains of a former world* (1804–11). *Journal of the Society for the Bibliography of Natural History* 7: 451–66.

Theunissen, Bert. 1986. The relevance of Cuvier's *lois zoologiques* for his palaeontological work. *Annals of Science* 43: 543–56.

Torrens, Hugh S., and John A. Cooper. 1986. George Fleming Richardson (1796–1848)—man of letters, lecturer and geological curator. *Geological Curator* 4: 249–72.

Twyman, Michael. 1970. *Lithography, 1800–1850: The techniques of drawing on stone in England and France and their application in works of topography*. London: Oxford University Press.

Wakeman, Geoffrey. 1973. *Victorian book illustration: The technical revolution*. Newton Abbot: David and Charles.

Wilcox, Donald J. 1987. *The measure of times past: Pre-Newtonian chronologies and the rhetoric of relative time*. Chicago: University of Chicago Press.

Williams, James. 1991. Dinosaurs—the first 150 years: A brief history of dinosaur discovery. *Geoscientist* 1 (4): 19–22.

Index

INDEX

INDEX